WHEN THE RIVERS RUN DRY

WHEN THE RIVERS RUN DRY

WATER—THE DEFINING CRISIS of the TWENTY-FIRST CENTURY

FRED PEARCE

FULLY REVISED AND UPDATED EDITION

BEACON PRESS
BOSTON

BEACON PRESS
Boston, Massachusetts
www.beacon.org

Beacon Press books
are published under the auspices of
the Unitarian Universalist Association of Congregations.

Printed in the United States of America
21 20 19 18   8 7 6 5 4 3 2 1

This book is printed on acid-free paper that meets the uncoated paper
ANSI/NISO specifications for permanence as revised in 1992.

Text design and composition by Kim Arney
Maps by Hardlines Studio, UK

Library of Congress CIP DATA is available.
ISBN 978-0-8070-5489-5 (pbk.) | 978-0-8070-5490-1 (ebook)

For Joe,
another river who died before his time

CONTENTS

ABBREVIATIONS

CGIAR	Consultative Group on International Agricultural Research
EEA	European Environment Agency
FAO	UN Food and Agriculture Organization
GERD	Grand Ethiopian Renaissance Dam
GIZ	Deutsche Gesellschaft für Internationale Zusammenarbeit (German government development agency)
IIASA	International Institute for Applied Systems Analysis
IIED	International Institute for Environment and Development
IPCC	Intergovernmental Panel on Climate Change
IS	Islamic State, also known as ISIS, ISIL, or Daesh
IUCN	International Union for Conservation of Nature
IWMI	International Water Management Institute
MDGs	Millennium Development Goals
OMVS	Organisation pour la Mise en Valeur du Fleuve Senegal (Senegal River Basin Development Organization)
TNC	The Nature Conservancy
UNEP	United Nations Environment Programme
UNESCO	United Nations Educational, Scientific, and Cultural Organization
UNICEF	United Nations Children's Fund
WHO	World Health Organization
WWF	World Wildlife Fund

INTRODUCTION

The Power of a River

IT WOULD BE wrong to say that I have been obsessed with rivers all my life. But they have a hold on me. I was brought up in a village in southeastern England where two small rivers begin their journey to the sea. The Stour flows east through the cathedral city of Canterbury, takes in some extra water bubbling up from the East Kent coal mines, and finishes in a marshy mess called Pegwell Bay, near the ancient Cinque Port of Sandwich. The Len flows west to the Medway and the Thames Estuary. It once powered eighteen mills on its ten-mile journey, filled the moat of a twelfth-century castle where Henry VIII often stayed with his Spanish first wife, Catherine of Aragon, and formed a lake in a park where I played school rugby and cricket. Then one day it flooded and I couldn't get to school. That struck me: the power of a river unleashed.

Rivers often define our world. Is there a better way of seeing London than taking a boat down the Thames to Greenwich? Is there a better book about America than the one narrating Huckleberry Finn's journey on the Mississippi? Some of the greatest human adventures have been along rivers: up the Orinoco to find El Dorado, or the search for the source of the Nile. Millions of Indians keep bottles of holy Ganges water in their homes. We romance on the Blue Danube and the Seine, and fight over the Jordan and the rivers of Babylon. The Yellow River is China's "joy and sorrow."

Yet something disturbing has been happening to our rivers. I have only slowly become aware of it: just a news story here and there. But the maps in my atlas no longer seem to accord with reality. The old geography lessons about how rivers emerge from mountains, gather water from tributaries, meander down floodplains, and finally disgorge their bloated flows into the oceans are now often fiction. Many rivers are dying as they go on, not growing. Inland seas, lakes, and wetlands that rely on them are disappearing too.

Because I am a journalist, I assembled a small file of press clippings. The Nile in Egypt, the Yellow River in China, the Indus in Pakistan, the Colorado and Rio Grande in the United States—all were reported to be trickling into the sand, sometimes hundreds of miles from the sea. Individually, these were interesting stories. Taken together, they seemed to be something more. Some kind of cataclysm was striking the world's rivers. So began the idea for this book.

I soon learned more. Israel is draining the Jordan River into a giant pipe before it reaches the country that bears its name. There have been droughts on the Ganges, because India has sucked up the holy river's entire dry-season flow. The ancient river Oxus, once the Nile of Central Asia, has been diverted into the desert, leaving the Aral Sea to dry out. This was a real shocker; the shoreline still shown on many maps of what was once the world's fourth-largest inland sea is hundreds of miles distant from where the shore is today. Back home, even the chalk streams of my childhood are disappearing.

Wells have been drying up too. More than half a century of pumping water from beneath the Great Plains of the United States has removed underground water that it will take the rains two thousand years to replace. In India, farmers whose fathers lifted water from open wells with a bucket, now sink boreholes more than half a mile into the rocks—and still they often find no water. In less than half a century, Saudi Arabia has almost pumped dry one of the three largest underground water reserves on earth.

My book is a journey on the world's rivers: a journey of discovery to find out why we face this crisis, what happens when great rivers die, where we could be headed—and how we can restore the rivers' health and our hydrological future. Although it is mainly about rivers, it is also about how we use water: about the staggering amount that it takes to feed and clothe us, and about how the world trade in food, cotton, and much more is also a trade in "virtual water"—the water it takes to grow those crops. That implicates Western consumers directly in the emptying of many of the world's great rivers. Water is a local resource, but the huge virtual-water trade has globalized the impact of emptying rivers.

As I took this journey, I met many water custodians. I shivered with men in overcoats who huddled in drafty offices trying to keep the Ruritanian badlands of Karakalpakstan from turning to desert. I marched in Japan with a man who once ran the world's largest dam-building organization but now campaigns to tear down dams. I met Indian rainwater harvesters and the last person alive with the secrets of divining water from under the hills of Cyprus. I drank toasts with Chinese bureaucrats who told me late in the evening how droughts on the Yellow River could one day trigger a flood disaster on a par with the one eighty years ago that they try not to talk about—it was man-made and killed almost a million people.

Here too are the stories of Muammar Gaddafi's Great Man-Made River, built to pump the world's greatest reservoir of freshwater from beneath the Sahara; of the river that flows backward to feed ten million people; of the extraordinary ancient water tunnels beneath Iran that could stretch two-thirds of the way to the moon; and the secret truth of the world's biggest-ever water-poisoning scandal.

~~~

In the decade since this book was first published, I have undertaken a host more journeys and fully updated the stories of earlier travels. In places, I have found that genuine efforts are now being made to revive some rivers. The Rhine in Europe is being given back parts of its floodplain, for instance. I have witnessed the efforts of Oman to revive its "unfailing springs." I have clambered to a mountaintop with a fog collector on the Canary Islands and drunk the products of Singapore engineers putting treated toilet water into bottles. I have talked to brewers about reviving Andean mountain "sponges" and seen inspiring efforts to prevent the death of the river Murray in Australia.

Too often these remain token efforts. A highly publicized "pulse" of water down the Colorado to replenish its delta lasted just eight weeks. In eastern Europe and much of Africa, I have seen a new urge to barricade natural rivers in the name of economic development. China has become the world's leading dam builder, its engineers blocking rivers such as the Mekong as they cross borders into neighboring lands, while building huge dams for willing customers across the planet. African governments continue to preside over the disappearance of lakes and wetlands on which millions of their poorest citizens depend.

In the course of this book, I hope to have answered some pressing questions. Can we fill the world's faucets without emptying its rivers? Do we, as engineers propose, need megaprojects to empty the Congo River into the Sahara, or divert the torrents of the Himalayas into the arid lands of southern India? Can we bring back the Aral Sea? Must the world's last wild rivers be dammed to generate hydroelectricity to fight climate change? Or should we think small, catching the rain from our roofs, irrigating crops with plastic sheaths bought from ice-cream salesmen, and saving rivers by turning our poo into fertilizer?

Nothing, perhaps not even climate change, will matter more to humanity's future over the next century than the fate of our rivers. Plenty of explorers have sought the source of the world's great rivers. My journey began as a mission to chart their deaths. But it is a hopeful journey nonetheless. I am an optimist. Water, after all, is the ultimate renewable resource.

I

When the rivers run dry . . .

the poor take to the road

1

THE HUMAN SPONGE

FEW OF US REALIZE how much water it takes to get us through the day. On average, we drink no more than a gallon and a half of the stuff. After including water for washing and flushing the toilet, we use only about 40 gallons each. In some countries, suburban lawn sprinklers, swimming pools, and sundry outdoor uses can double that figure. Typical per capita water use in suburban Australia is about 90 gallons, and in the United States, 100 gallons. There are exceptions, though. One suburban household in Orange County, Florida, was billed for 4.1 million gallons in a single year, or more than 10,400 gallons a day. Nobody knows how they got through that much.

We can all save water in the home. But as laudable as it is to take a shower rather than a bath, buy a low-flush toilet, and turn off the faucet while brushing our teeth, we shouldn't get hold of the idea that regular domestic water use is what is really emptying the world's rivers. Manufacturing the goods that we fill our homes with consumes a certain amount, but that's not the real story either. It is only when we add in the water needed to grow what we eat and drink that the numbers really begin to soar.[1]

The numbers are mind-boggling. It takes between 250 and 650 gallons of water to grow a pound of rice. That is more water than many households use in a week. For just a bag of rice. It takes 130 gallons to grow a pound of wheat and 65 gallons for a pound of potatoes. When you start feeding grain to livestock for animal products such as meat and milk, the numbers become yet more startling. It takes 3,000 gallons to grow the feed for enough cow to make a hamburger and between 500 and 1,000 gallons for that cow to fill its udders with a quart of milk. Cheese? That takes about 650 gallons for a pound of cheddar or Brie or Camembert. If you think your virtual-water shopping cart is getting a little bulky at this point, maybe you should leave that 1-pound box of sugar on the shelf. It took up to 400 gallons to produce. And the 1-pound jar of coffee tips the scales at 2,650 gallons—or 10 tons—of water. Imagine taking *that* home from the store.

Turn these statistics into meal portions and you come up with more than 25 gallons for a portion of rice, 40 gallons for the bread in a sandwich or a serving of toast, 130 gallons for a two-egg omelet or a mixed salad, 265 gallons for a glass of milk, 400 gallons for an ice cream, 530 gallons for a pork chop, 800 gallons for a hamburger, and 1,320 gallons for a small steak. If you have a sweet tooth, so much the worse: every teaspoonful of sugar in your coffee requires 50 cups of water to grow. Which is a lot, but not as much as the 37 gallons of water (or 592 cups) needed to grow the coffee itself. Prefer alcohol? A glass of wine or beer with dinner requires another 66 gallons, and a glass of brandy afterward takes a staggering 530 gallons.

At present, human societies globally consume an estimated 8.7 billion acre-feet of water every year from rivers and underground aquifers.[2] (An acre-foot is the amount of water it takes to cover an acre of land to a depth of a foot. For those of a metric disposition, it works out at a bit over a thousand cubic meters.) Divided among the planet's 7.3 billion people, that is 1.2 acre-feet each per year, or around 4 tons a day. Westerners consume more, of course. I figure that as a typical meat-eating, beer-swilling, milk-guzzling Westerner, I consume as much as a hundred times my own weight in water every day, or between 360,000 and 480,000 gallons a year. Hats off, then, to my vegan daughter, who gets by with less than half that. It's time, surely, to go out and preach the gospel of water conservation. But don't buy one of those jokey T-shirts with slogans like save water, bathe with a friend. Good message, but you could fill roughly twenty-five bathtubs with the water it takes to grow the cotton needed to make the shirt.

Where does all that water come from? In England, where I live, most homegrown crops are watered by rain. Even so, a lot of the food consumed in Britain, and all the cotton, is imported. In much of the world, the water to grow crops is collected from rivers or pumped from underground. Its diversion to fields is increasingly likely to deprive people downstream of water and to empty rivers and underground water reserves.

I checked out the geography of my water footprint. The corn in my breakfast cereal probably comes from the American Midwest, where underground water is drying up fast. Increasing amounts of my meat come from American cattle raised in sheds, where they are fed with irrigated alfalfa. Even chickens raised in England are fed on soy from Latin America. My sugar most likely comes from Cambodia or South Africa. The cotton in my jeans was grown in Uzbekistan, where it will have helped empty the Aral Sea.

Two-thirds of all the water that humans take from nature is used for agriculture. By some estimates, as much as 10 percent of that water—about 570 million acre-feet—ends up in international trade. The water is not itself traded, of course; the products of its use are. Economists sometimes call this "virtual water." As countries across the world run short of water, this trade in the virtual stuff is increasingly contentious.

The biggest net exporter of virtual water is the United States. It sends abroad in traded goods around a third of all the water it withdraws from the natural environment. Much of that is in grains, either directly or via meat. The trade is emptying critical underground water reserves, such as those beneath the High Plains. The United States exports an amazing 80 million acre-feet of virtual water in beef. Other major exporters of virtual water include Canada (grain), Australia (cotton and sugar), Argentina (beef), and Thailand (rice). Israel and arid southern Spain both export water in tomatoes, Ethiopia in coffee.

Cotton-growing countries provide a vivid example of this perverse water trade. Cotton grows best in hot lands with year-round sun: deserts, in other words. Old European colonies and protectorates such as Egypt, Sudan, and Pakistan still empty the Nile and the Indus for cotton growing, as they did when Britain ruled and Lancashire cotton mills had to be supplied. When the Soviet Union transformed the deserts of Central Asia into a vast cotton plantation, it sowed the seeds of the destruction of the Aral Sea. Commissar cotton is still in business in today's Uzbekistan. Much of the water from the shriveling sea has in effect been exported over the past half-century in the form of virtual water.

By far the biggest importing region is the European Union. It sucks in some 260 million acre-feet of other countries' water, to meet 40 percent of its virtual-water needs.[3] Britain imports 38 million acre-feet a year, the equivalent of more than half the flow of the river Nile. Some 70 percent of the water needed to feed and clothe the country comes from beyond its borders, a percentage only exceeded in the Middle East and by Switzerland, the Netherlands, Belgium, and Malta.

Nobody should want to ban the virtual-water trade. Iran, Egypt, Jordan, and Algeria would starve without it. "The Middle East ran out of water some years ago. It is the first major region to do so in the history of the world," says Tony Allan of King's College London, who coined the term "virtual water."

Without the trade there would have been continual water wars in the region.[4] Economists say the trade should significantly reduce global water needs for growing crops, because it can allow their cultivation where water requirements are less.[5]

That's the theory. However, for many crops, such as cotton and sugar, the trade in virtual water looks like terribly bad business for the exporters. Pakistan consumes more than 40 million acre-feet of water a year from the Indus River—almost a third of the river's total flow—to grow cotton. Is that prudent? And what logic is there in the United States mining the High Plains aquifers to add to a global grain glut? Whatever the virtues and vices of the global trade in virtual water, the practice lies at the heart of some of the most intractable hydrological crises on the planet.

2

LAKE CHAD

Wetlands to Killing Fields

ON LAKE PUNJAMA the waters were still high, and hundreds of birds were swooping low. Locals took me out in a flat-bottomed boat. We landed some catfish. In the soft early-evening light, we rowed past boys cutting the tall grasses to use as floats for their fishing nets. Nearer the shore, two brothers were wading waist-deep, driving their cattle to the same grasses to feed. The fishers shouted to the cattle herders to keep their animals clear of the fishing lines, which were spread out across the water.

That was the scene when I first visited the Hadejia-Nguru wetland in northern Nigeria, in 1992. It was a dramatic, kite-shaped expanse of green and blue stretching for more than 60 miles along the southern edge of the Sahara Desert, a bulwark against the advancing sands. Its lakes were full of fish, and every wet season its fertile waters overflowed across the land, creating lush pastures that sustained tens of thousands of cattle and watered more than half a million acres of fields that farmers planted with crops as the floodwaters receded. All told, the wetland supported a million or more people and provided exports of fish and vegetables to cities across the most populous country in Africa.[1]

It appeared idyllic, but all was not what it seemed. After we brought our boat ashore on an island in the lake, I met a family of Fulani cattle herders setting up camp for the night. Like other Fulani, they came to the wetland during the dry season to find water and graze their animals on the lake edges and on stubble in fields after the harvest. The woman rounded up the cattle, shooed them into a stockade, and lit a fire of dung and straw to keep away the mosquitoes. But her husband was morose. "Once many Fulani were coming here," he said. "There was more water then, and more grazing land. Now there is little grazing. We still come, because there is nowhere else to go, but there are always disputes with the farmers." The previous year there had been pitched battles between the Fulani and gangs of settled farmers, he said. By the time the

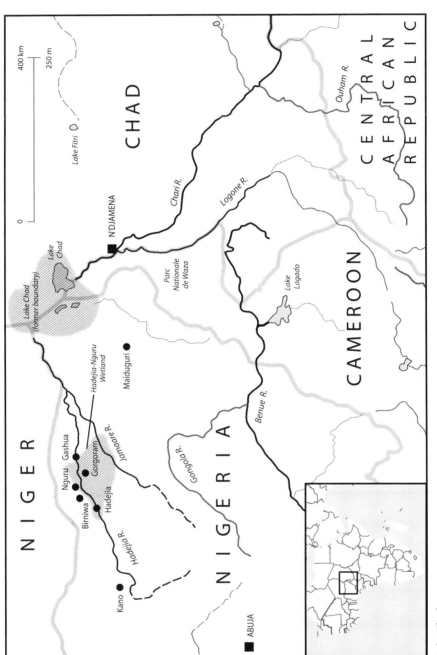

Lake Chad

fighting died down, bodies were floating in the water and ten people were dead. "Every year, people get killed."

I moved on to Gorgoram, once a sizable town in the heart of the wetland. It was holding its annual fishing festival. Teams of youths had come from hundreds of miles around to compete for the biggest catches in the last pools as the flood retreated. Dignitaries from across northern Nigeria came too, awarding prizes and lobbying for votes in upcoming elections. By 1992, the festival was failing, however, because the floods were poor and there were few fish. I watched as they hauled in desultory catches that barely moved the scales during the judging. Gorgoram was emptying. And all around, the landscape was littered with fallen trees, victims of a declining water table. On the road from Gorgoram to Nguru town, we drove through a desiccated wasteland that went on for miles, until suddenly a man appeared from behind one of the last trees left standing. He was like an apparition, and probably sensed it. He advanced on our car. "This is how men are suffering in the bush," he roared and would say no more.[2]

The next morning, in Gashua, the easternmost town on the wetland, fish were in short supply in the market. The traders said the catches were smaller these days. What the market stalls did have in abundance were bags of potash—the salty crust of desert soils, scraped by villagers from shriveling oases to the north. It was the one local commodity in plentiful supply.

Later, we passed a camel train—a dozen animals tramping through the desert bearing potash to the market. The animals had just crossed the bed of a dried-up river. It had been like that, locals said, since a big dam, the Tiga Dam, was finished upstream two decades before. Camels were walking where fishermen once threw their nets. Leading the animals was a bearded man in white robes, carrying a sword. His piercing blue eyes glared at us as he walked on. I have always remembered him. He seemed to me at the time like an omen for worsening and more threatening times. It felt like the desert was on the march and with it, its people.

My destination that day was Maiduguri, a market town that trades the produce of the Lake Chad region, of which the wetland was a part. Once, it would have been a lake port. The lake has long since retreated, however. I was told that in its efforts to keep going, Maiduguri had already exhausted two aquifers and was pumping from a third. Officials from a European Union antidesertification project in the town told me that without a return of the lake, this final aquifer would never recharge. The town too would die.

A quarter century after my visit, Maiduguri has not quite died. But those parched communities in which I had sensed a rising tide of anger are now the heartland of the Islamic extremist group Boko Haram. The town and its hinterland regularly make the news as the refuge of suicide bombers, kidnappers of schoolgirls, and the murderers of farmers in their fields. Journalists now visit only under armed guard.[3]

The Hadejia-Nguru wetland was once a complex network of natural channels and lakes that mingled the waters of the Hadejia River, coming in from the west, and the Jama'are, from the south. Since the Nigerian government built two dams on tributaries of the Hadejia, 80 percent of that river's annual flow has been captured. The government promised that the dams would turn the landscape green and create wealth. The Tiga Dam, completed in 1974, supplied a state irrigation project south of Kano, the biggest city in northern Nigeria. Then, shortly after my visit, the Challawa Gorge Dam began to hold back water to supply the French-built 30,000-acre Hadejia Valley irrigation project.

The two dams destroyed a natural bounty of far greater economic as well as ecological value than all the crops grown in these irrigation projects. Today, four-fifths of the wetland has been lost. Lakes like the Punjama where I caught my catfish have died. And without water percolating down from these wetlands into the soil, the water table has declined by as much as 80 feet, unleashing a further round of desiccation.[4] Meanwhile, on the Jama'are River sits the half-completed Kafin Zaki Dam. It would dry out the remainder of the wetland for good. The money for this ran out long ago, but local politicians still want to finish it.

It is too easy to see communities that depend on natural wild resources and the vagaries of untamed rivers as backward and left behind by progress. The truth, quite often, is the opposite. It is they who know the truth about how to make the maximum use of their natural resources. It is the urban sophisticates with their engineering degrees who haven't got a clue. The fate of the Hadejia-Nguru wetland tells that story well. Economists have concluded that the big dams and irrigation canals upstream of the region have been a huge waste of money. The value of the crops grown on the irrigation projects is tiny compared with the lost benefits on the wetland from farming, fishing, harvesting timber, raising livestock, gathering honey, cutting straw for construction, and making bricks.

Edward Barbier of Colorado State University has studied the economics of Hadejia-Nguru, totting up the economic gains and losses from the dams. There have been winners and losers, he says. The losers predominate, however.

"Gains in irrigation values account for at most around 17 percent of the re-sulting losses on the floodplain"—$68 per acre lost against $12 gained.[5] The unfinished projects, he says, should be abandoned. The only sensible use for the existing dams is to re-create the natural flood. Sadly, the bigwigs in Nigeria have yet to take such advice.

The death of a wetland is a terrible thing, particularly a wetland in a desert. When it happens, lakes shrivel, crops go brown in the baking sun, fishing nets empty, trees crash to the ground, and herders slaughter their animals for what-ever pitiful amount of cash they can raise. The local people depart the badlands, while bad people move in. Often this happens when the rains fail. How much more terrible when the drought results from human action—when the wetland dies because humans have decided to divert the rivers that should replenish it; when the water is taken for little human benefit, often as a statement of the power of one community over another; and when even hard-nosed economists say that what is going on is madness. This is happening all across the Sahel region of Africa, the border zone between the Sahara desert to the north and farmlands to the south. For the Sahel, seasonal lakes and natural wetlands such as Hadejia-Nguru are the lifeblood. But the lifeblood is failing, and nowhere is this more devastating than around Lake Chad.

The lake is a shallow inland sea into which several major rivers drain. Its catchment stretches across eight countries, from Algeria to Sudan. The lake has always been vulnerable to changes in the flow from those rivers. It is rarely more than 13 feet deep. Since evaporation can remove more than 6 feet of water from its surface in a year, it needs constant supplies of water or it dies.[6]

In good times, Lake Chad spreads across up to 10,000 square miles. With diminishing inflows, it has lost more than 90 percent of its surface area since the early 1970s. Once, four countries bordered the lake: Nigeria, Niger, Chad, and Cameroon; now only the last two have banks on its shores or access to its waters. What remains of the lake is mostly covered with reeds and small islands.[7]

Initially the lake's decline was largely due to long droughts in the 1970s and 1980s.[8] Since then, however, there has been a general rising trend in rainfall, while the lake has remained stubbornly empty.[9] The reason is that the rivers that once brought water to the lake have been captured. Most of the waters from the wetter south of the lake's basin, flowing out of northeastern Nigeria, and down the rivers Chari and Logone in Cameroon, are now diverted. Irrigation projects conceived during the drought years were aimed at delivering economic development. While the water has benefited those able to secure irrigated land, people downstream all the way to Lake Chad have been left high and dry.[10]

The Logone flows out of the rain-soaked fringes of the great Congo rainforest of Central Africa. Its waters had for centuries sustained a thriving floodplain rich in wildlife and with ample pickings for fishermen and hunters. Farmers planted traditional varieties of rice on the wet soils left by the retreating floods each year. It was one of the great centers of the Fulani, the largest nomadic pastoral group in the world.

In 1979 a state-owned rice company built the Maga Dam to divert its headwaters in the Mandara Mountains to irrigate new paddy fields created on the floodplain. The company also raised 60 miles of embankments on the floodplain to prevent unwanted floods from the river. The result of all this engineering was to deprive the river of water, to reduce its floodplain from more than 3,000 square miles to less than 160 square miles, and to cut the flow into Lake Chad by 55 percent. Some twenty thousand head of cattle had to move away. Fish yields fell by 90 percent. Output of the local varieties of sorghum and rice fell by 75 percent. Water tables fell, emptying wells and water holes. Elephants and lions in the adjacent Waza National Park fled one of their last refuges in Central and West Africa. A hundred thousand people who had depended on the floods for their livelihoods were left destitute.[11]

To compound the folly, the rice project was spectacularly inefficient even on its own terms, says Paul Loth, now of Leiden University and an expert on the river. Little rice was grown most years, and only around a tenth of the water collected by the dam was productively used. Like the irrigation projects in northern Nigeria, the project "diminished rather than improved the living standards and economy of the region as a whole," he told me. "A once fertile floodplain was turned into a dust bowl."

The catastrophe has extended to Lake Chad, for which the Logone had been the biggest source of water. As the lake retreated, some thirteen million farmers, fishermen, and herders in the region suffered water shortages, crop failures, livestock deaths, collapsed fisheries, soil salinity, and increasing poverty.[12] The result has been social breakdown and conflict. As a Nigerian government audit of the Lake Chad Basin in 2015 found: "Uncoordinated upstream water impounding and withdrawal, land degradation, soil erosion, deforestation and bush burning . . . has created high competition for scarce water, resulting in conflicts and forced migration."[13]

This has culminated in the growing tyranny of Boko Haram. The brutal Islamist insurgency, which is based in northern Nigeria, has been killing and abducting people around Lake Chad. Between 2013 and 2016 more than 2.3 million people were displaced as a result of the environmental and social chaos, including 1.3 million children.[14] Many joined the million migrants who entered Europe during that time, causing political angst across the continent. European Union politicians called for an effort to address the "root causes" of this exodus. What were they? Poverty, civil conflicts, and the spread of terror groups have all been blamed. Yet among people living in the Lake Chad Basin, few doubt that much of this exodus has arisen from the degradation of its rivers and wetlands and the drying up of the lake itself.

"Youths in the Lake Chad Basin are joining the Boko Haram terrorist groups because of lack of jobs and difficult economic conditions resulting from the drying up of the lake and extinction of its resources," Mana Boukary of the Lake Chad Basin Commission, an intergovernmental body, told the international broadcasting network Deutsche Welle.[15] Toby Lanzer, the United Nations humanitarian coordinator for the Sahel region, agrees. "Asylum seeking, the refugee crisis, the environmental crisis, the instability that extremists sow—all of those issues converge in the Lake Chad basin."[16]

Towns in northeastern Nigeria, in the Hadejia-Nguru wetland, in Maiduguri, and down the dried-up rivers to the lake, have been especially badly hit. After massacres in 2015, the citizens of Doron Baga, a former fishing port near the lake, fled en masse to refugee camps over the border in Chad. There was no safety there, however. In late 2015, suicide bombers targeted the fish market and refugee camp in the town of Baga Sola, and the Chad government declared a state of emergency in the region around the lake.[17] Months later, the bodies of forty-two Cameroonian, Nigerian, and Chadian fishermen, killed by Boko Haram, were pulled from Lake Chad in Cameroon.[18]

Without a revival of the lake, rivers, and wetland ecosystems, the region may have entered a downward spiral of environmental decay, economic decline, breakdown of law and order, and continued massive outward migration. It's what happens when rivers run dry.

3

SAHEL

Tragedy of the Floodplains

IN SENEGAL, YOUNG MIGRANTS declare they are headed for "Barca ou barzakh"—Barcelona or death. A journey I took in mid-2016 along the valley of the river Senegal, which forms the border between Senegal and Mauritania, revealed that its wetlands, once a source of natural wealth, have lost their fecundity. A reengineering of the river carried out in the name of economic development has had the opposite effect. Wetlands along the river are drying out—and farmers, herders, and fishers are leaving.

Seydou Ibrahima Ly, a teacher in the riverside village of Donaye Taredji, told me: "Compared to the past, there aren't many fish. Our grandparents did a lot of fishing, but we don't." He blamed the Manantali Dam hundreds of miles upstream. "In the past, the river had a flood that watered wetlands where fish grew. Now there is no flood because of the dam." It was the main reason why around one hundred villagers have left, he said. "In some villages, they are almost all gone."

"We have seen an increase in numbers of migrants leaving the country," said Nicolas Mendy of the Senegalese Red Cross, based in Saint-Louis, the pretty former capital of French West Africa at the mouth of the river Senegal. I met him during a break in a meeting with families of migrants who had disappeared on the perilous journey to Europe. "They mostly go to Mali or Niger by bus, then cross the Sahara to Libya and take boats to Europe," he said.

Since the 1970s drought in the Sahel, the nations of the Senegal Basin have sought economic development by harnessing the river's flow to generate hydroelectricity and irrigate fields along its banks. On behalf of member governments, the Senegal River Basin Development Organization (OMVS) commissioned two dams. The Manantali in upstream Mali held back floods and created a regular year-round flow; the Diama in the river's large delta prevented salt water from the Atlantic Ocean penetrating upstream.

The dams have benefited some residents, especially those gaining hydro-electricity. They have created havoc too. The Manantali Dam ended the seasonal flood on which millions of farmers depended to water their riverside fields. After its completion, there were land disputes on the border between Senegal and Mauritania that escalated into a full-blown war between the two countries.[1]

The problems did not stop when the war ended. The OMVS conceded in an environmental analysis of the basin that eliminating the river's annual flood has created extensive ecological problems—and human losers as well as winners. The project "has made flood-recession crops and fishing on the floodplain more precarious, which makes the rural production systems of the middle valley less diversified, and therefore more vulnerable."[2]

The dams and irrigation schemes on the Senegal River have destroyed 90 percent of the fisheries in the river's delta. They have also caused the loss of up to 600,000 acres of seasonally flooded land, where farmers planted crops as the waters receded, pastoralists grazed their animals, and forests and wildlife flourished.[3] The UN Environment Programme has reported a "severe reduction in natural pasture areas" with "access to the river . . . very difficult for cattle," resulting in "frequent conflicts between stockbreeders and farmers" along the river.[4] When I asked Amadou Lamine Ndiaye, the director for environment at OMVS, why this was allowed to happen, he shrugged his shoulders and told me: "The agency was created to build dams for irrigation and hydropower and to regulate flows; it has no fisheries, pasture, or ecosystems mandate."

In the middle reaches of the river, I found that many villages are emptying. After visiting Donaye Taredji, I drove to nearby Guiya, which has suffered what local farmer Idrissa Fall told me was a "mass exodus that began in the late 1980s." He had nearly joined them, "but my mother told me not to." Twenty years later, he had no regrets but said that young people who did leave home returned a few years later far richer than he could ever hope to be at home.

"The migrants know the boats [across the Mediterranean] are dangerous, but they have a determination to go and find a better life," said Oumar Cire Ly, deputy chief of the neighboring Donaye village. "My elder brother went to France and he is now a teacher at the University of Le Havre," he said. "He still sends money back to the village." Cire didn't expect the flow of migrants to halt any time soon.

Some of the worst problems are in the river's coastal delta. Large areas are drying out, and channels and lakes clog with silt and fast-growing invasive weeds. The 30-mile-long Lac de Guiers, once the largest lake in the delta, is overgrown with weeds. The infestation happened after water was diverted for a giant sugar farm and into a pipeline to supply the nation's capital, Dakar, 130 miles to the south. The 116,000-acre Ndiael wetland was once a haven for pelicans, manatees, white pelicans, endemic mud turtles, and around a hundred species of fish. It simultaneously provided grazing for tens of thousands of cattle. All that ended when the channel that provided it with water dried up.[5]

Then, an Italian company, Senhuile, owned by the Italian Tampieri Financial Group, promised to revive the wetland by digging a canal from Lac de Guiers. In return it would take some of the water for a farm it wanted to build on the northern part of the wetland, where it would irrigate biofuel crops. The government backed the scheme. By 2014, when I visited, the water was passing to the company's fields but without reaching the wetland. At the wetland, there was a new lookout tower so visitors could watch the abundant bird life. But the water channel to it was clogged with weeds. The wetland was bone-dry and, needless to say, there were no birds to watch.

Afterwards, out in the midday sun, I witnessed a two-hour confrontation between villagers and Senhuile's head of security and spokesman on conservation, Colonel Ibrahima Diop. The colonel angrily insisted that he and his officials "have done a lot for the community," and he promised the water would soon come. The chief of Daymane, one of the local cattle-herding villages, did not believe him. It was "a new form of colonization," he told me. "The company has made us prisoners in our own land."

The Senegal government has a target to expand irrigation from the current 250,000 acres to an eventual 1 million acres, much of it in the Senegal River valley. If that policy comes to fruition, then most expect the result will be less water in the river and further damage to wetlands, their ecosystems, and local economies. There will be fewer fish, fewer pastures, less flood-recession agriculture, more poverty among the millions who depend on wetland resources— and more angry scenes like the one I witnessed. No doubt there will be more migrants too.

$$\approx\!\approx$$

The Inner Niger Delta in northern Mali is another green oasis in the Sahel. Covering an area the size of Belgium, it is the largest floodplain in West Africa. It has the legendary ancient city of Timbuktu on its northern shore and, in the fifteenth century, was the heart of the Empire of Mali. These days, however, it feels like a backwater ignored by the Mali government in faraway Bamako.

The delta forms where West Africa's largest river, the Niger, spreads out across flat desert land. Its lakes and wetlands expand and contract with the seasons, depending on flow in the river. In summer, when it rains in the highlands of Guinea, the flow is strong and the lakes and wetlands of the delta fill with water and fish. I was able to travel by boat across a waterscape, passing hundreds of fishing boats and huge flocks of birds visiting from Europe. We docked at villages that were islands amid the flood. Many of their inhabitants were away fishing the waters.

During the long dry season the waters recede. Travel is by four-wheel drive. But in many low-lying areas, farmers move onto the damp and fertile sediments left by the receding waters. They plant rice, millet, and other crops. Meanwhile, cattle herders bring their animals from across the region to feed on the rich grasses exposed by the retreating water.

Fishers, farmers, and herders from different ethnic groups have traditionally cooperated to make the maximum use of the delta's resources. At its largest, the delta covers just 1.6 percent of Mali, but it provides pasture for a third of the country's cattle, delivers 8 percent of its GDP, and sustains two million people, 14 percent of the population. Its fish harvest is exported across West Africa.[6]

All this is threatened by the Markala Barrage, built on the river Niger just upstream of the delta. It diverts water for irrigating thirsty crops like rice and cotton in giant state-sanctioned farms. Abstractions of water from the river at the barrage have cut the area of delta flooded each year by up to 7 percent.[7] Less water has meant fewer forests, fisheries, and grasses. It has led to disputes between herders and farmers over the available wetlands and pastures. In 2012, when Islamist groups exploiting the tensions invaded the delta, many of the area's inhabitants fled.[8] There was a military intervention from the French, the old colonial power in the region. A peace deal followed, but tensions remain high. In early 2015, a new jihadi organization called the Macina Liberation Movement, made up largely of Fulani pastoralists, began launching attacks on farmers in the delta. (*Macina* is the Fulani term referring to the delta region.)[9]

In the past, the Inner Niger Delta has been a refuge in times of drought and trouble elsewhere in the Sahel. Now it is becoming a source of conflict and outmigration. People-smugglers follow old salt-traders' tracks across the desert to the Mediterranean. Journalists report many Malians among those in boats heading from Libya to Italy.[10]

Many more may follow. Upstream in Guinea, the government in 2017 announced the go-ahead for a new hydroelectric dam to be built by China. It would hold back the river's wet-season flow to generate year-round power. The Mali government supports the plan because it wants to use the regular river flow from the dam to triple the area of land irrigated by the river. Everyone seems to have forgotten about the delta and the natural wealth that its annual flood provides. Hydrologists estimate that the combined impact of the government schemes will cut both fish catches and cattle pastures by 30 percent. The new hydrological normal would be similar to the conditions during the last major drought here in 1984, when three-quarters of the delta dried out and a million people fled.[11]

4

RIDING THE WATER CYCLE

EARTH IS THE WATER PLANET. It contains an unimaginable 1.1 quadrillion acre-feet of the stuff. But more than 97 percent of that is seawater, which humans cannot drink and cannot, except in very local circumstances, afford to purify. "Water, water everywhere, / Nor any drop to drink," as Coleridge's Ancient Mariner put it. Of the remaining 28 trillion acre-feet of freshwater on and near the planet's surface, two-thirds is locked up in ice caps and glaciers and one-third, about 9.7 trillion acre-feet, is in liquid form. The greater part of this fresh liquid water is in the pores of rocks. These reservoirs of underground water, known as aquifers, vary widely in their accessibility and drinkability. But the water is there, beneath our feet.

The remaining smidgeon of the world's liquid freshwater—we are now down to a mere 162 billion acre-feet or so—is above ground. The biggest volumes, around 71 billion acre-feet, are in lakes, and there are probably another 71 billion acre-feet in soils and permafrost. Next comes atmospheric water vapor, which contains another 10.5 billion acre-feet, or about a thousandth of 1 percent of all the water on earth. After that come swamps and wetlands at 9 billion acre-feet; rivers, which contain around 1.6 billion acre-feet at any one time; and living organisms, from rainforests to you and me, with about 800 million acre-feet.[1]

This is a static picture: where the water is at any one moment in time. Water is not a static resource, however. It is constantly on the move, through soils and into ancient geological formations, down rivers, into the ocean depths, freezing and melting, evaporating into the air and forming clouds to fall again as rain. If we are concerned about how much water is available to us in a form that we can keep on using, the static figures don't tell us much. It is the dimensions of these movements, known as the water cycle, that are critical.

Take aquifers. The crude statistics suggest that these underground water reserves should easily be our number one resource. And yes, the volume of water down there is huge, dependable, and accessible, ready to be pumped to the surface. With current technology, perhaps a tenth of the known resource could be extracted. Sink a well and away you go. But as everyone knows, wells run dry. Pumping out an aquifer will empty it forever. Unless that aquifer is being refilled, we are mining water. So a more useful measure of the water available to us underground, year in and year out, is how much of that water is replenished by rain.

Looked at this way, underground water shrinks in importance. By some perversity of nature, many of the biggest aquifers are beneath deserts. They were filled in wetter times but are now "fossil" water resources, with no prospect of recharge from rains if we pump them out. The largest quantities sit in the pores of sandstone rocks beneath the Sahara and, until recently, the Arabian Peninsula. Other big reserves are beneath the Australian outback and the arid High Plains of the American Midwest. We can and do pump this water, but as we do, the water level falls, pumping costs rise, and quite often water quality deteriorates.

This is not just a problem for desert regions. An analysis published in 2017, after the sampling of some ten thousand wells worldwide, reckoned that most of the world's water more than 650 feet underground is fossil water that is essentially nonrenewable.[2]

Water in the other big reservoirs—the oceans and ice caps—has a similar long residence time. But if most of the water in what we might call the "slow" water cycle moves at a snail's pace, water in other parts of the water cycle—generally the parts with smaller volumes at any one time—moves around quickly. This "fast" water cycle comprises water that is constantly on the move: evaporating from land and sea, forming water vapor and then clouds, raining to the earth and then evaporating again.

Some water from the fast cycle disappears into the slow cycle. Rain percolates into rocks or gets frozen on glaciers or trapped in inland seas or tied up in plants and animals. Some water moves from the slow to the fast cycle—in meltwater from glaciers, or bubbling from rocks in springs, for instance. But in a year, about 400 billion acre-feet of water flows through the fast water cycle. In practice, there is much less water; it just circulates several times. If this entire annual cycle were collected together, it would cover the entire earth's surface to a depth of almost 3 feet.

Because most of the planet is covered by ocean, the larger part of this fast cycle is again of not much use to us. It comprises water evaporating from the sea and falling back onto the sea in rain. For the rest, we are beginning to home in on the part of the cycle from which we get most of our water. First there is the water that falls as rain onto the land and then evaporates from the land again. Some evaporates from soil; some is taken up by plants as they grow and is then released from leaves in a process called transpiration; some evaporates from bodies of water, including artificial reservoirs. Second, there is a constant flow of water from the land, where rainfall exceeds evaporation, to the sea, where evaporation exceeds rainfall. About 32 billion acre-feet of water makes the journey from the land to the sea every year. Or it did before we started diverting it. Most of the water we use around the world today comes from this final part of the water cycle.

So how much of this runoff do we use? Of those 32 billion acre-feet a year, much hurtles off the land in occasional floods or flows away from permanent rivers. Of the river flows, hydrologists estimate the maximum that might reasonably be caught and used by humans employing current technology is 11 billion acre-feet. However, nature has played one more trick on us. Many of the world's greatest rivers are in regions where few people can or want to live. The three rivers with the biggest flows—the Amazon, the Congo, and the Orinoco—all pass through often-inhospitable jungle for most of their journey from headwaters to the sea. Those three rivers alone carry almost a quarter of the water we have to survive on. Two more of the top ten—the Lena and the Yenisei, in Siberia—run mostly through Arctic wastes. Take out these and we are left with around 7 billion acre-feet of river water for our annual needs, or about 1 acre-foot per person on the planet, of which we already extract approaching half.

Not bad, but earlier I calculated my own annual water use at between 360,000 and 480,000 gallons a year, which works out to between 1.1 and 1.5 acre-feet. I imagine most of the world would like to live as well as I do. So we have an emerging problem.

Water is heavy stuff. It's not easy to move around unless gravity is with you, as anyone who has carried a full bucket any distance will agree. So all these global numbers can still give a misleading impression of how much water we can really get our hands on. The water business, like the property business, comes down to a matter of "location, location, location." So where is the water?

Just six countries have half of the world's total renewable freshwater supply on their territory—Brazil, Russia, Canada, Indonesia, China, and Colombia. Some smaller countries have a greater amount of water available per capita of population. Greenland's sixty thousand citizens have more water than anyone else. Each of them could make use of 8 million gallons of it every day without having to defrost a single piece of ice. But having no crops to irrigate, they need little. Alaskans have available 1 million gallons a day each. Also weighing in above 100,000 gallons each a day are the citizens of the Congo, Iceland, and the three neighboring South American rainforest states of Guyana, Suriname, and French Guiana.

Meanwhile, people in the driest countries have by far the greatest need—to irrigate crops as well as to quench their thirst. The Palestinian desert enclave of the Gaza Strip is the most water-starved political unit on earth, with just 37 gallons of brackish underground water a day for each inhabitant. Others at the bottom of the hydrological heap include small desert states such as Kuwait and the United Arab Emirates, and island states like the Bahamas, the Maldives, and Malta.

Even at the continental level there are clear haves and have-nots. Europe and North America are well served—especially considering that irrigation demands are mostly fairly low. Australia is the driest continent but has a small population. Asia, by contrast, has almost two-thirds of the world's population but only one-third of its runoff, 80 percent of which is concentrated in the short monsoon season. Africa's continent-wide allocation isn't too bad, but a third of its runoff goes down a single river, the Congo. South America has only 5 percent of the world's population but glories in three of the world's top ten rivers—the Amazon, the Orinoco, and the Parana. Fully a quarter of the world's runoff occurs here, though little of this water is near people. The vast and wet Amazon basin, with 15 percent of the entire world's runoff, has just 0.4 percent of the world's population.

Even within countries, water can be inconveniently distributed in time and place. Most of India gets all its rain in a hundred hours during a hundred days. Parts of Ethiopia suffer perennially from drought and famine, even though 84 percent of the flow of the Nile River begins within the country's borders. Canada has 90 percent of its water where 10 percent of its people are, and vice versa. If northern China were a separate country, it would be one of the most water-stressed in the world.

All this has come to matter a very great deal because of two changes over the past half century: the soaring world population, and the manner in which we have gone about trying to feed that population. Half a century ago, the world was gripped by Malthusian nightmares of mass starvation. Women gave birth to an average of five children each. The global population was set to double in a generation. In his influential book *The Population Bomb,* the biologist Paul Ehrlich declared that "in the 1970s hundreds of millions of people are going to starve to death . . . the battle to feed all of humanity is over." As famines spread through Africa in the early 1970s, he was not alone in holding such apocalyptic views.[3]

It didn't happen that way. The global population has since doubled, but scientists defied the doomsayers by breeding new high-yielding varieties of rice, wheat, and corn that kept the granaries full. Henry Kissinger was proved right when he declared at a World Food Conference in 1974 that "for the first time, we may have the technical capacity to free mankind from the scourge of hunger." He misjudged the world's ability to get food to those who needed it, so his follow-up—that "within a decade no child will go to bed hungry"—was sadly wide of the mark. Africa in particular has missed out. But, averaged globally, the "green revolution" has kept growth in food production ahead of growth in population.

While the new crops delivered more grains per acre of land, however, they often delivered less grain per acre-foot of water. The green revolution has kept the world's granaries full, but it has emptied the rivers. The new high-yield varieties often produce less "crop per drop" than the varieties they replaced. As a result, the world today grows twice as much food as it did a generation ago, but it abstracts three times more water from rivers and underground aquifers to do it. In arid countries such as Egypt, Mexico, Pakistan, and Australia, and across Central Asia, 90 percent or more of the water abstracted from the environment is for irrigation. In the green-revolution countries, water consumption per capita is several times that of European countries. Pakistan abstracts five times more water per person than Ireland does, Egypt five times more than Britain, and Mexico five times more than Denmark. The United Nations' Food and Agriculture Organization says that on at least a third of the world's fields today, "it is water not land that is the binding constraint" on production.[4]

No wonder the rivers are emptying. As they do, farmers are increasingly taking the capture of water into their own hands. A generation ago, government agencies were overwhelmingly in charge of how water was abstracted from nature, transported, and used. They built the dams, dug the canals, and managed the irrigation systems. But in the past two decades across the poor world, hundreds of millions of small farmers have been turning their backs on governments and finding their own solutions to their water problems. They can buy cheap pumps and other low-tech and off-the-shelf equipment to extract underground water. They have been capturing rainfall and storing water on their land. "We are amazed at the scale of what is going on," says Meredith Giordano of the International Water Management Institute (IWMI), author of a study on the phenomenon.[5]

Rich farmers have always had the money to buy their own pumps, but "with the availability of cheap Chinese pumps, this type of irrigation is accessible to a much larger range of farmers," agrees Charlotte de Fraiture of UNESCO's Institute for Water Education in Delft, the Netherlands. "The capacity of even a small pump with one to five horsepower is bigger than most farmers need. So they hire them out when they don't need them."

In India, small-time rural entrepreneurs travel the countryside on bikes or donkey carts, with pumps strapped on the back. They rent the pumps for a dollar an hour, so even the poorest farmers can get some water from a local river or aquifer. Governments and donors alike "continue to focus their attention and investments on the underperforming public irrigation sector, when private irrigation is both more important and has larger potential" for scaling up, says de Fraiture.[6]

Too often, we see poor smallholder farmers as passive victims of natural disasters, or the grateful recipients of aid. In this narrative, they emerge in a different light. It is they—rather than governments, nongovernmental organizations (NGOs), or Western aid-givers—who are the active players, taking charge of their own destiny. There is a new democracy about water today that some see as a hydrological liberation but others see as anarchy. For there is a downside, which in some regions is a major threat to both future water supplies and the survival of the farming communities themselves.

Independent action by farmers to water their fields is sometimes creating a "tragedy of the commons"—in which everyone grabs what water they can while they can, because they know that all will suffer when the water runs out. This is especially a risk where farmers are pumping out underground water reserves at rates that the rains cannot replenish. The nightmare scenario now is that underground water reserves will dry up. For while rivers are constantly refreshed by the rains, many aquifers will never recover.

In 2016, Arjen Hoekstra of the University of Twente assessed the world's water insecurity.[7] He found that four billion people were affected by water scarcity for at least one month a year. Water scarcity didn't necessarily mean people literally had nothing to drink. But they were overusing the water resources at their disposal, so risked running out in the longer term. "Groundwater levels are falling, lakes are drying up, less water is flowing into rivers, and water supplies for industry and farmers are threatened," he concluded. The future capacity of the world to water its crops is itself in jeopardy. [8]

There is one last threat we need to consider: salt. Salt is the unnoticed poisoner of the world's farmland. Every day, 5,000 acres of agricultural soil become unusable because of the damage caused by salt. It affects more than a fifth of the world's irrigated soils, according to a 2014 assessment from United Nations University (UNU). [9]

The salt comes from farmers irrigating their fields. The salt is in the water. It dissolves from rocks as rivers flow from mountains. In the rivers, the concentrations of salt are tiny. The problem is that it concentrates in irrigated fields, because when the water is drained away, evaporates, or is incorporated into crops, most of the salt stays behind. It accumulates in the soil and eventually, in the worst cases, a bright white crust forms on the surface of irrigated fields, which become toxic and infertile. Fields turn to salty desert.

As artificial irrigation has grown in importance, the buildup of salt is becoming a huge and growing problem. Irrigated fields produce more than a third of the world's food—including the majority of food in some of the most populous countries, such as Pakistan, China, India, and Indonesia. Projects that initially greened the desert are now creating desert.

The total area affected by salt has risen from around 100 million acres in the early 1990s to 150 million acres today, an area the size of France. The UNU study put the annual economic cost of lost crop production at $27.3 billion. This cannot continue, said lead author Manzoor Qadir. Rehabilitation is difficult. Options include treatment with chemicals such as gypsum, improved field drains, and flushing soils of salt before each planting season. All are expensive, and extra water for flushing is often simply not available. But with world demand for food set to increase by two-thirds by midcentury, preventing losses to salt and bringing salt-damaged farmland back into use should become a global priority.

As artificial irrigation has grown in importance, the buildup of salt is becoming a huge and growing problem. Irrigated fields produce more than a third of the world's food—including the majority of food in some of the most populous countries, such as Pakistan, China, India, and Indonesia.

II

When the rivers run dry . . .

America loses the West

5

RIO GRANDE

Crossing the River

THEY SERVE A STRONG BREW at the Alamo coffeehouse in Presidio, a small farming town on the United States–Mexico border. They need to. Times are tough, said Terry Bishop, looking up from his second mug. This land, next to the Rio Grande in Texas, had probably been continuously farmed for longer than anywhere in America, he said. Six hundred years, at least. It had been home to scalp hunters and a penal colony; it had seen Comanche raids, Spanish missionaries, marauding Mexican revolutionaries, and a population boom during past amnesties for undocumented immigrants. All that time it had been farmed. But soon, he said, his land would be back to sagebrush and salt cedar.

After a short ride in his pickup truck, we climbed the levee by the river at the end of his last field, and Bishop showed me the problem. The once mighty Rio Grande was reduced to a sluggish brown trickle. Here in its middle stretches, the river often dried up entirely in the summer. All the water had been taken out by cities and farmers upstream. "The river's been disappearing since the fifties," said Bishop, who remembered the wetter times when his father ran the farm. For 200 miles upstream of Presidio, there was no proper channel anymore, he said. That was why they called it "the forgotten river."

Bishop's land brings with it legal rights to 8,000 acre-feet of water a year from the river—enough to flood his fields to a depth of more than 3 feet and grow almost any crop he wants. In recent years he had taken only a quarter of that, however. Yields got so low, the farm went bust. The whole area had gone bust. Presidio was once a major farming center. It used to ship in thousands of Mexican workers to harvest its crops. The Bishop farm alone once employed a thousand people. But that has all ended, and the unemployment rate among the town's permanent residents is almost 40 percent.

About the only profitable business was desert tourism, he said. An old silver mine a few miles up the road had been turned into a "ghost town," and a

Rio Grande

fort at Cibolo Creek is now an upmarket resort where Mick Jagger once stayed. Harvesting tourists, that's the game now.

We were speaking during a drought across the American West that lasted for more than a decade and peaked in 2013. Not every year was bad, but when water did flow down the Rio Grande, reservoirs upstream in New Mexico were grabbing all they could, and the river downstream mostly remained dry.

A decade on, in 2017, I checked out the state of affairs down in Presidio. After heavy snows in the Colorado Mountains the previous winter, they had a brief respite. Climatologists were predicting that the letup would be short-lived, however. Megadroughts could last for decades as global warming intensified. Bishop still grew a little alfalfa, a grain that is fed to cattle. But he had just turned over many of his fields to create a natural wetland, supplied with waste-water from a local sewage-treatment plant. It was the only water on offer. On his wetland, he had planted bulrushes, cottonwoods, and willows. Within days, avocets, stilts, and mallards had returned.[1] The farmer had become a wildlife conservationist. Times are changing down on the Rio Grande.

The Rio Grande stretches almost 2,000 miles from the snowfields of the Colorado Rockies to the Gulf of Mexico, via New Mexico and Texas. It drains 336,000 square miles, a tenth of the continental United States and more than two-fifths of Mexico, where it is called the Rio Bravo. For 1,200 miles, it forms the boundary between the US and Mexico. Much of it flows through the Chihuahuan Desert, one of the most biologically diverse deserts in the world, with jaguars, wolves, falcons, and prairie dogs.

The hub of human exploitation of the Rio Grande is the Elephant Butte Reservoir, just upstream of El Paso, Texas. It was built in 1915 and changed the river forever. The wild, untamed flow had been home to sturgeon and eels, though it also obliterated villages and once rode right through downtown Albuquerque. All that was ended for good by the dam, after which the river's waters were corralled for irrigation. Today, the dam all but empties the upper river to supply local farmers and El Paso. More than nine million people in the basin rely on the Rio Grande's waters, but most of its water is taken to grow two of the thirstiest crops in the world, cotton and alfalfa. The wastage is huge. Evaporation in the hot sun takes more than 3 feet of water a year from the Elephant Butte, a total of up to 130,000 acre-feet. That represents a loss of up to 16 percent of the water put into supply.[2] Taking account of field losses, only about 40 percent of the water captured from the river reaches the crops.

I moved on to El Paso, where the Chamizal National Memorial commemorates a treaty that fixed the boundary between El Paso and its Mexican twin city, Juárez, by forcing the meandering river to pass down an unchanging concrete canal. This brutal carve-up may have underlined the river's political importance, but it hardly accorded it respect. Today the river is virtually invisible from the memorial behind a high chain-link fence designed to keep out illegal Mexican immigrants. Only up on the ugly, heavily guarded border bridge can you see it—a fetid trickle in an absurdly wide concrete canal flanked by a six-lane highway and a container dump. There was so little flow that as I watched, the wind ripping upstream was blowing the water back toward distant Colorado.

El Paso is in hydrological trouble. With the river now trickling through the town virtually empty, the city relies heavily on groundwater in drought years. It has been buying up properties from local farmers to bag their rights to underground water reserves. Its "water ranches" are dotted all along the highway to Presidio. Tiny cattle towns like Valentine and Dell City keep El Paso's faucets full, for now at least, for water ranching is only a temporary solution. The major aquifer here—which underlies 15 million acres of Texas, New Mexico, and the Mexican Chihuahuan Desert—is itself running dry, because there isn't enough rain to replace what is pumped out.

The *El Paso Times* has for years regularly alerted readers to the days when they can use public water on lawns, and the days they can't. Jittery suburbanites are repairing old wells in the hope of capturing some private water from beneath their land.[3] In some of the unplanned shantytown *colonias* where Mexicans usually end up after crossing the river, thousands of people still live without any access to piped water—and that is shocking to find in the United States in the second decade of the twenty-first century, even in the desert.[4]

Downstream of El Paso, the river becomes a dribble of sewage effluent that disappears into remote scrub. Hydrologically speaking, the river that rolls out of the Colorado snowpack and once ransacked Albuquerque dies now amid the saline pools of the forgotten river. The only water for the channel downstream of Presidio comes from Mexico. Back in Presidio, standing on the levee on Terry Bishop's land, I could see the empty Rio Grande bed being filled from the south by the Rio Conchos. For much of the time, the Rio Grande downstream of Presidio is the Rio Conchos.

I was intrigued about how that happened. How come the US sucks up all the water from the Rio Grande only for the Mexicans to replenish it? The answer lies in the terms of the treaty, signed in 1944, that shares out the waters of the Rio Grande and its tributaries. The treaty makes no requirement on the

US to give water to Mexico, but it does require that one-third of the average amount of water flowing into the Rio Grande from six Mexican tributaries, much the largest of which is the Rio Conchos, is allocated to the United States. The trouble for the Mexicans is that however low the tributaries flow and however little rain falls in the deserts of northern Mexico, the Mexicans are still required to deliver annually for US use a minimum of 350,000 acre-feet, averaged over five years. The United States can dry up the Rio Grande at El Paso as much as it wants, but come hell or low water, the Mexicans have to deliver that quota.

By and large, they do. At the start of this century, Mexico was for a while more than four years in arrears. Things got acrimonious, with Texas farmers from Presidio to the Gulf of Mexico demanding their water rights. Eventually, Mexico paid up. At the end of the most recent five-year cycle, in late 2015, Mexico was 263,000 acre-feet in debt, or almost a full year's worth. By the end of January 2016, it had paid in full.[5]

Thunder clapped in the canyons and it started to rain as I headed over the border to Presidio's twin town, Ojinaga, on the Rio Conchos. Long before the arrival of Europeans, Native Americans lived along the Rio Conchos, hunting the animals that congregated beside its waters. Early Spanish explorers came north along the river, before moving on into Texas and New Mexico. A century ago, Ojinaga was the headquarters for Pancho Villa's Mexican revolution.

The Ojinaga Irrigation District had been a source of prosperity here for a generation, taking water from the Rio Conchos. The irrigated area covered 22,000 acres, growing cotton and corn in the Chihuahuan Desert. But here, as in Presidio, farming had subsided for want of water. Under the shade of cottonwood trees on his small farm, I met sixty-eight-year-old Rito Guerrero. He came down from the Sierra hills to farm here when he was young and the irrigation canals were new. "Once, the water came right up from the river. All my land would be watered," he told me. He still got water sometimes from the canals that run right past his gate. "But it's six years since I planted corn," he said. "I tried watermelons two years ago, but I lost everything. Cotton had a good market, but there isn't enough water. I could be living in the USA with my son."[6]

Farmer-cum-teacher Humberto Lujan took me around the irrigation district. Most of the canals were dry and their sluice gates closed and rusting. Three-quarters of the fields were abandoned. The sagebrush was returning. "Virtually nothing comes down the river anymore," he said as we peered into the town's empty reservoir and spotted the invasion of salt cedar along the riverbank. The town had lost a quarter of its population. "The young have gone north to the States, and the old are selling up," he explained.

Northern Mexico is littered with places like this. I took the bus south through the desert to the Delicias Irrigation District, which stretches for more than 60 miles east of the town of Chihuahua. It is the largest irrigation district on the Rio Conchos. Its glory days were in the 1960s and 1970s, when farmers became rich growing cotton, which they called "white gold." Then it became the cornerstone of a big dairy industry. The cattle fed on alfalfa grown on its irrigated fields.

Most of the fields are fed from water captured from the river at the Boquilla and Madero Dams. The Madero Dam is decorated with two statues, one carrying corn and the other cotton. It has a recreation area with water slides and advertises itself as a fishing resort. The initial effect is impressive but misleading. When I visited, the reservoir was at a third of capacity. Beneath a bridge over the 1,300-foot-wide riverbed, the stream of water was only about 6 feet wide. Of the district's ten thousand farms, only about a third could be sure of getting any water most years. And none got water in winter as they once did. "Some areas have not been irrigated for ten years," said the district engineer Ezequil Bueno Torres.

During my visit, the district was engaged in a big modernization process to save water.[7] It was lining irrigation canals to prevent water from seeping through their porous bottoms. The $130 million scheme was paid for mostly by the US government, and the aim was to save 285,000 acre-feet of water annually. When it was done, the farmers all assured me, there would be more water for everyone, including the Americans downstream. That, of course, was why the Americans were funding it.

There was a fallacy here, however. The water they were "saving" was not actually wasted at all. The farmers had for years been pumping the seepage water up from underground. Many relied on it to water their crops. At the time of my visit, this recycled water accounted for about a fifth of all the irrigation in Delicias. If the seepage was cut, their wells would dry up. I asked about this. The local irrigation engineers, and the people funding their work, seem unconcerned by this prospect. The tragedy is that to meet their immediate obligations to deliver water to Texas farmers, the Mexicans are imperiling the long-term future of their underground water reserves. The win-win was set to turn into a lose-lose, as the ratchet emptying the Rio Grande was given another turn, this time in the name of efficiency.

6

COLORADO

A Busted Flush

SAN LUIS RÍO COLORADO sits on the banks of the Colorado River, where the river passes beneath a bridge and across the border from the United States to Mexico, headed for the ocean. Most years, no water passes this way anymore. Then, one day in March 2014, a small miracle happened. The river resumed its former flow down its dried-up bed and briefly replenished the long-parched delta 30 miles farther south on the shore of the Gulf of California. A delta landscape where jaguars and beavers once roamed had not seen fresh river water since 1993.

The pulse of water was a result of water being released upstream, out of Lake Mead, the giant reservoir behind the Hoover Dam, and allowed to flow on through the Morelos Dam, the last in the Colorado River system. For half a century, these dams had kept the water back to feed canals that irrigate desert fields and keep the faucets flowing in Southern California, Nevada, and Arizona. Just for a few weeks, the engineers on the river let nature have a share. At San Luis Río Colorado, a party was held on the riverbank to celebrate.[1]

This symbolic rewetting was just an experiment—to find out what would happen on the delta. It was encouraging. Native willows and cottonwoods germinated, and a flush of green spread. But afterwards, the delta was again dry. The bloom was soon gone.[2]

It is a truism that water won the West. The 1,450-mile Colorado, which drains a twelfth of the continental United States, has become the lifeblood of seven states, delivering its water to growing cities, feeding irrigation projects, and generating hydroelectricity.

Two giant reservoirs control the flow of the middle reaches of the Colorado and insure supply to the lower states. The first, Lake Mead, was filled in

Colorado River

the 1930s behind the Hoover Dam in Boulder Canyon. Then, in 1964, the Glen Canyon Dam drowned a series of spectacular gorges to create Lake Powell, which was named after John Wesley Powell, a one-armed Civil War hero who in 1869 made the first boat journey by a white man down the river.[3]

The two reservoirs collect water when snowmelt in the Rocky Mountains fills the river and distribute it to cities and fields during the long summer growing season. Having more than four times the capacity of the river's average annual flow, they can also even out fluctuations between wet and dry years. While urban areas are taking an increasing amount, most of the water abstracted from the river still goes to irrigating some 4 million acres of fields in the river valley and in Arizona and California.

America has always subsidized farming in the West, and today perhaps $1 billion a year is still poured into keeping farmers irrigating crops that they would not otherwise grow. Subsidies encourage waste. Every year several million acre-feet of water evaporates from reservoirs, farm ponds, and flooded fields, while much of what does get to crop roots is used to grow low-value crops such as alfalfa. Even in dry years, the presumption is that a wet year will be along soon. Recently, that presumption has looked increasingly foolhardy.

The Colorado is both legally and hydrologically one of the most regulated rivers in the world. But it is becoming clear that the legal and the hydrological no longer mesh. A century ago, more than 20 million acre-feet of water flowed unimpeded to the Gulf of California every year. When lawyers shared out the river's waters between the states in 1922, on the eve of the dam-building era, they gave 7.5 million acre-feet to the upper basin states of Colorado, Utah, Wyoming, and New Mexico and another 7.5 million acre-feet to the downstream states of California, Arizona, and Nevada. With another 1.5 million acre-feet assigned to Mexico, that added up to around 16.5 million acre-feet.

That should have left water to spare, but ever since, flows have been diminishing. Between 2000 and 2014 the average flow was just 12.4 million acre-feet, "the worst 15-year drought on record," according to Jonathan Overpeck of the University of Arizona. Most of the decline seemed to be due to global warming; continued droughts could be expected.[4]

Nobody is happy and something will have to give. Upstream, the state of Colorado is not happy at being required to export three-quarters of the snowmelt from its mountains when its own farms and cities are running low. Denver and Colorado Springs don't quite see why they have to shut off their sprinklers so Phoenix, Las Vegas, and Los Angeles can keep theirs on.

In 2016, the water in Lake Mead, the largest reservoir in the US, was at its lowest level since it was first filled in the 1930s. The Bureau of Reclamation, its custodian, said the reservoir was at just 37 percent of capacity. (This can't be blamed on the brief flood-pulse release the year before, which had taken just 1 percent of the river's annual flow.) A good snowfall in the mountains in the winter of 2017 brought some relief.[5] Even so, at average river flows, hydrologists said it would take the reservoir ten years to refill; and there were few grounds for thinking even average flows were likely.[6]

In the twenty-first century, Lake Mead has typically been overdrawn by 1.2 million acre-feet a year, enough to cause its level to fall by 12 feet a year. Clearly this cannot continue. California, Nevada, and Arizona, the last three states to receive the river's water, have a combined population of thirty million people dependent on the river's water, plus a million acres of irrigated farmland and turbines that need water to generate hydroelectricity. In 2007, the three states reached an agreement that when the level falls to an elevation of 1,075 feet above sea level, they will begin making reductions in their abstractions. California, for instance, would have to take a cut of 200,000 acre-feet a year; Arizona would lose 320,000 acre-feet. That could happen as soon as 2019. Steeper cuts would follow, if reservoir levels continued to fall. These states are living on borrowed water and borrowed time.

What can be done? Growing thirsty crops like cotton and alfalfa is clearly crazy. But, with an increasing proportion of the Colorado's water already going to cities, there is greater pressure on urbanites to moderate their water use too. Some cities are taking up the challenge. Most surprising, perhaps, is Las Vegas. To many, it is the world capital of ostentatious consumption, a place that wastes water in the desert with as much disregard for tomorrow as its visitors waste money. Well, not quite. Not anymore. Behind the showy glitz of its ludicrous hotel fountains, Vegas today claims to be a model of water conservation.

It's not that the city's water users are skimping, exactly. Bylaws allow more than 6 feet of water on golf courses in a year; fountains and lakes are unimpeded. Some of the water savings, moreover, reflect nothing more than changing national standards. For instance, the amount of water used to flush the toilet has been cut across the country, thanks to bigger valves and S-bends that allow a shorter, faster, more efficient flush. The standard flush is now 1.6 gallons rather than 3.4 gallons. Even so, since 2000, Vegas has increased its population by 34 percent while cutting its water use by 26 percent. That is impressive.

Things seem different 300 miles away in another shining desert city, Phoenix. With a population now well past three million, Phoenix has in recent decades been one of the fastest-growing urban areas in the US. It sprawls across the Sonoran Desert with no apparent limit. In the mid-1990s, it was growing by more than two acres every hour. It is scarcely less now. Metropolitan Phoenix covers nearly 400 square miles so far, and it has been as profligate with water as with land. Savings made in its toilet cisterns have been squandered with outside use of water.

The residents of private estates in smart suburbs such as Scottsdale and Paradise Valley water their lawns and fill their swimming pools as if they lived anywhere but a desert. Private lakes are blossoming. Developers compete to offer the greenest golf courses, the most luxuriant gardens, and the tallest fountains. Fountain Hills, one of the outer suburbs, installed what was for a while the world's highest-shooting fountain, spurting 600 feet into the air for thirty minutes of every hour throughout the day. Half of the city's water is used for "landscaping." It is the main reason that Phoenix contrives to use a third more water per head than the national average. Through the summer, residences in some suburbs consume around 1,000 gallons of water a day.[7]

There used to be a river flowing through Phoenix—the Salt River. In pre-Columbian times, Hohokam Indians used it to irrigate fields. It was, according to archaeologists, the largest prehistoric irrigation system in North America. Dams have long since dried up the river, however. The red-light district was built across its old bed. Phoenix and its suburbs grew increasingly reliant on water pumped from beneath the desert. But water tables have been falling fast. Despite groundwater management laws passed back in 1980, Arizona pumps up more than twice as much as the sparse rains can replenish. Arizona mines water like no other region in the United States.

The end of Phoenix's profligate water use would have come by now if it had not been for the city fathers, who as early as the 1950s began lobbying Washington for cash to bring in water from out of state. Arizona saw its future in taking water from the great overused waterway of the American West, the Colorado River. After twenty-five years of persuasion and the influence of an Arizona senator as chairman of the Senate Appropriations Committee, pork-barrel politics delivered the goods, and in the 1980s, the Bureau of Reclamation built for Arizona one of the world's largest and most expensive water-delivery systems. The Central Arizona Project is capable of taking almost 1.6 million acre-feet a year out of the Colorado River and pours it into a concrete canal 300 miles long that zigzags across the desert to Phoenix and its smaller sister, Tucson.

The project cost $3.6 billion to build and another fortune to run, and it loses 7 percent of the flow to evaporation en route.

Arizona today gets more than a third of its water from the giant canal. Its hydrological bounty unleashed the latest real estate boom in Phoenix. Fearing the good times may not last, Arizona has for years been taking more than it needs and storing the rest in recharged underground aquifers. It has banked about 9 million acre-feet in all.[8] The bank will soon have to start paying out, however. Recent economies aimed at protecting Lake Mead, and forestalling further draconian reductions, have reduced the Central Arizona Project's take from the Colorado to around 1.3 million acre-feet a year. That is still more than a tenth of the river's total average flow, and much more in dry years.[9] So Phoenix is starting to rethink its absurd water demands. There is talk of the state paying for lawn "buy-back," meaning residents in the opulent suburbs will be paid to dig up their thirsty ornamental turf. But if the megadrought continues, then further steep cuts in supply are inevitable. There is just not enough water to sustain Arizona's Sun Belt lifestyle.

If water shortages don't put the farmers out of business, salt may be the apocalypse awaiting the great agrarian civilization of the American West. As Arthur Pillsbury, the doyen of hydrologists studying the region, said before his death in 1991, "The Colorado basin will eventually become salt-encrusted and barren because of salt." The only question is when.

All down the river, more and more salt is clogging up the system. It is flowing downstream from the headwaters in the Rockies. It is also being washed from soils and bedrock in irrigated areas like Paradox Valley in Colorado and Wellton-Mohawk in Arizona. Almost all the water flowing down the Colorado leaves the river several times to irrigate fields and returns via drains. At each step, it both loses volume, through evaporation, and picks up salt. So the concentration of salt in the water increases as it travels downstream. At its headwaters, the Colorado contains about 50 parts per million of salt. By the time it reaches the Hoover Dam, near Las Vegas, it contains more than 700 parts per million. The river and the extensive artificial irrigation and drainage networks that circulate its waters have become a vast system for collecting and distributing salt. Each year about 10 million tons of salt enter the system, but virtually none reaches the ocean. Tens of millions of dollars are spent on farms every year trying to minimize the problem, but even so, annual crop losses from salt are currently estimated at $330 million.

Outside Yuma, just upstream of San Luis Río Colorado and the border with Mexico, is the ultimate technical solution to the salt problem. There lies a giant, $300 million desalination plant. It was built a quarter century ago to clean up the briny flow from the Wellton-Mohawk Irrigation and Drainage District, the last in the United States, as it returns to the river close to the border. The idea was to deliver the cleaned-up flow to Mexico as part of the US treaty obligation to supply 1.5 million acre-feet of usable water annually for irrigation over the border. Apart from a couple of pilot runs, the Yuma plant has never been used.[10] The desalted water would cost ten times more to produce than its value to farmers on either side of the border. So the plant remains as a macabre memorial to the grand designs of ever-greater use of the Colorado's water. If Pillsbury is right, it may be there still on the banks of the Colorado, long after the fields whose drainage waters it was intended to cleanse have become choked with salt.

7

CALIFORNIA

Running on Empty

THEY HAVE BEEN MEASURING the snowpack at Phillips, California, in the Sierra Nevada, on the first day of April every year since 1941. That day is chosen because it is the point at which the winter buildup of snow should have peaked but before serious melting begins. Typically, the researchers' dipstick finds 5 feet of snow. But on April 1, 2015, for the first time in seventy-four years, there was no snow—only dry earth. The state's governor, Jerry Brown, still made the journey for the annual measurement. But only so he could stand at the snowless spot and declare that the state was in a hydrological crisis. With no snow, there would be no spring melt and the rivers would be empty for much of the summer. He imposed a 25 percent cut in water use across the state.[1]

California has been running out of water for years. Half the state is desert, and for the rest, precipitation, whether rain or snow, just hasn't been what it was. January 2015 was its driest January since records began over a century ago. The Sierra Nevada normally provides a third of the state's water. But it was in its fifth year of drought—perhaps the worst spell for a thousand years. Combined with higher temperatures, which meant more winter precipitation falling as rain, the drought meant that the snowpack—the state's most reliable source of water—was faltering.[2]

Other California water sources were doing no better. Underground water was being pumped dry, especially in the agricultural heartlands of the Central Valley, where farmers had always been allowed to pump as much as they wanted—with predictable consequences. The state's pumping has for many years exceeded the replenishment from rainfall by about 15 percent. That meant an overdraft of 1.3 million acre-feet a year.

The Colorado River, which supplies much of Southern California, had also been suffering from more than a decade of low flow. Its two great reservoirs, Lake Powell and Lake Mead, were at their lowest levels since the Hoover Dam was built on the river more than eighty years ago. The day was drawing ever closer when the diversions from the river to California, along with Nevada and Arizona, would have to be reduced. California enacted laws intended eventually to end overpumping of groundwater—at the expense of large areas of irrigated farmland. Water agencies gave rebates for people digging up their lawns and installing low-flush toilets.[3]

The drought broke in 2017. Suddenly there was snow again in the Sierra Nevada. By spring, reservoirs were overflowing. Hundreds of thousands were evacuated as others threatened to burst. In Kern County, deep in Central Valley, there were floods. Several people, unused to the idea of keeping out of the way of water, drowned. Water managers saw an opportunity, however. They ramped up their efforts to capture some of the floodwater and secrete it underground in storage aquifers, especially in the normally parched Central Valley.

The problem, they swiftly discovered, was that during the long drought, many aquifers had lost their storage capacity. Without water, the porous honeycombs of rocks had collapsed under the weight of the rocks above. At the surface, the collapsing rocks have caused roads to crack and pipelines to fracture, but underground, the space available for storing water has declined. One estimate is that this crushing has permanently reduced the water storage capacity of the Central Valley aquifers by half—the equivalent of blowing up a dam the size of Grand Coulee.

This is inconvenient, to say the least. California's hydrologists predict that persistent megadroughts are going to be a feature of climate change across the American West. Megadroughts punctuated by occasional flashes of extreme wet weather.[4] California, with the West's largest population and the country's biggest agricultural output from irrigated fields, will likely suffer the worst. "We are in a new era," Governor Brown told reporters, as he surveyed the snowless scene at Phillips. "The idea of your nice little green lawn getting watered every day, those days are past." Golf courses and lawn sprinklers were a particular target in his crackdown, with large farms in the state also forced to cut water use. Water, they used to say out West, flows uphill to money. No longer, it seems.

Nowhere better epitomizes the water crisis in Southern California than the Salton Sea, sitting in the desert close to the Mexican border. In truth, there should be no sea here. But there has been for over a century now, though for how much longer remains unclear.

The Salton Sea was all the fault of Charles Rockwood, a land speculator in California's boom years at the start of the twentieth century. He and his buddy George Chaffey, who had already made a fortune planting orange groves in Los Angeles, dreamed of turning a desert depression close to the Mexican border into an agricultural boom town. They planned to do it by capturing some of the flow of the mighty Colorado River as it passed 60-odd miles to the east at Yuma, Arizona.[5]

Despite the distance, achieving the diversion didn't look too difficult. The bottom of the depression they had earmarked for irrigation was 230 feet below sea level, the second-lowest point in the whole United States. It was downhill most of the way. For part of this distance there was an old riverbed through which to channel the flow. So, once the Colorado's west bank had been breached at Yuma, the water would come. In 1901, Rockwood's California Development Company constructed a rickety wooden dam and canal, and began tapping some of the Colorado's flow.

To attract farmers, Rockwood and Chaffey did a swift bit of rebranding. They had picked up their desert depression at a rock-bottom price partly because it was known to locals as the "valley of death." It sounded much better when renamed the Imperial Valley. The soils were good, too, and settlers eager to make their fortune came thick and fast. Within four years, some fourteen thousand of them had staked out farms, dug irrigation canals from the new desert canal, and planted 300,000 acres of fields irrigated with Colorado water. Traders moved in to service them. New towns such as Calexico, El Centro, and Brawley sprouted. It was like a gold rush.

It seemed too good to be true, and it was. For with the water came silt. The Colorado is the second siltiest river on earth, after the Yellow River, and by 1904, its muddy waters had clogged Rockwood's canal. The water dried up, and crops were dying in the Imperial Valley. Facing down angry mobs of farmers, Rockwood cut a deal with a landowner over the border in Mexico to dig another canal from the Colorado to Imperial Valley, down an old creek known as the Alamo. Then, disaster struck again. The following year, the Colorado was in full flow when it broke its banks near the new canal and began pouring into the desert. By the autumn, the entire flow of the Colorado—32 acre-feet a second—was coursing into the Imperial Valley. There, having nowhere else to go, it created an inland sea.

As the flood subsided, the river stuck with its new course. It seemed to have permanently given up its old route to the Pacific Ocean and settled on a one-way ticket into the desert. It set about remaking the region's physical geography to fit its needs. There was a small waterfall where the new Colorado entered the new lake. It soon began to cut back upstream, eating up the soft bed of the new river. Such was the power of the river that for a while the waterfall was retreating through the desert at a rate of more than half a mile a day. It gouged out a gorge 100 feet deep and 1,000 feet wide. It was as if a new Grand Canyon were being created before the world's eyes.

The federal government panicked. Farms were disappearing beneath the sea, and nobody was quite sure where the water would go when the river had filled the desert depression. There wasn't much time. So the government called in the Southern Pacific Railroad Company, which had already seen the flood consume a chunk of its line from Los Angeles to Tucson. The company spent eighteen months dumping six thousand railcar loads of rock, gravel, and clay into the desert to force the Colorado back onto its old course. Finally, on February 10, 1907, it succeeded. The Colorado resumed its old course south into Mexico and to the Gulf of California. But it left behind a new sea in the Imperial Valley covering some 600 square miles.

If all the farmers of the Imperial Valley had departed, the lake would probably have evaporated to nothing by now. The farmers were not done, however. They regrouped and lobbied for a more permanent canal to bring Colorado water to California. The All-American Canal, finally completed in 1938, was one of the first and largest major diversions of water from the Colorado. It was finished soon after the Hoover Dam, which helped regulate the wilder flows of the river and prevent future disasters. The canal has since become the basis for a billion-dollar business in the Imperial Valley, providing salad crops for sale across the country. The Imperial Irrigation District is today the largest single diverter of water from the Colorado River. The descendants and corporate successors of the farmers who stuck it out through the bad times have grown rich.

The sea is still there too, thanks to the large amounts of drainage water pouring from the farmers' fields. In its new role as a desert sump, it gradually became more polluted and saltier as pesticides and the salt brought down by the Colorado accumulated there. It soon became known as the Salton Sea. The federal authorities had the idea of filling it with fish. First they tried striped bass, then salmon, halibut, bonefish, anchovies, and turbot. All died. In the 1950s they tried thirty species of fish from the Gulf of California, and some of these, especially orangemouth corvina, did much better.

Despite the accumulating toxins and occasional epidemics of disease among the fish, the Salton Sea became one of the world's most productive lakes, home to an estimated two hundred million fish. The fish attract birds in large numbers. Strange to say, with the rest of California draining its wetlands for new real estate, the toxic sump became the state's premier bird habitat, providing a home for 380 species, including egrets, cormorants, brown pelicans, and various boobies. It is home to more species than the Florida Everglades, America's most famous wetland. It hasn't just attracted birds. In the 1960s, before Las Vegas emerged as a desert oasis for the jet set, the Salton Sea was the place to be seen. Frank Sinatra came with his Rat Pack, and yacht clubs, beauty contests, and nightclubs followed. Even the Beach Boys put in an appearance on a shorefront that had everything except surf.[6]

Water is a precious commodity in California, however, and water politics has always been central to its economy. Suffice to say that, somehow, the Imperial Valley's farmers became legally entitled to almost three-quarters of the state's share of the Colorado: an annual dose of some 3 million acre-feet of water entered their valley. For years, California took more than its legal entitlement of the Colorado's flow—5 million acre-feet a year, rather than the allowed 4.4 million acre-feet.

When the droughts started to hit, and with other states growing increasingly thirsty, the federal government ruled that enough was enough. On New Year's Day 2003, it cut flows to the Golden State back to the legal limit. By now, cities were needing ever more of the state's water. So to avoid having the cities run dry, the state decided to meet its new obligations by persuading the Imperial Valley farmers to give up almost a fifth of their share, so it could roll on down the All-American Canal to San Diego. The state agreed to make things right with the farmers by having San Diego pay for lining the canals that distribute water through the valley and investing in more efficient ways of irrigating, such as drip irrigation.

It seemed like a neat solution. But there was a problem, at least for the survival of the Salton Sea. The sea had persisted throughout the long decades because it was where the drainage water flowing off the fields of the Imperial Valley ended up. The inefficiency of the farmers' irrigation systems had been the sea's salvation. However, if the farmers poured less water onto their land and left more for San Diego, then less drainage water would flow off the fields and into the Salton Sea. With inputs expected to fall by at least a quarter, the legacy of Rockwood's rocky engineering a century ago would start to dry out. "On-farm conservation" would trigger "off-farm Armageddon," warned the then director of the Salton Sea Authority, Tom Kirk.

The authorities recognized the problem. During the first years of the new arrangement, the farmers were required to put extra water into the Salton Sea to compensate for the loss of farm runoff. The farmers did this by leaving parts of the irrigation district fallow. Even so, since 2005, the surface of the sea has fallen by about a half foot per year, exposing thousands of acres of former lake bed to the desert's blowing winds. Meanwhile, the requirement for compensation waters expired at the end of 2017.[7]

With no replacement water in sight, defenders of the sea have been warning afresh of impending disaster. Michael Cohen of the Pacific Institute, a water think tank, warned that from the end of 2017 the lake will start to rapidly diminish. By 2030 "the amount of water flowing into the lake will decrease by about 40 percent; its surface will drop by 20 feet, and its volume will decrease by more than 60 percent; salinity will triple; and the shrinking lake will expose 100 square miles of dust-generating lake bottom to the region's blowing winds, worsening the already poor air quality in the region."[8] In early 2017, the state unveiled plans for methods to suppress dust from the drying lake but not to replenish the lake itself.[9]

Meanwhile, with expectations growing that from 2019 or so, California's entitlement to water from Lake Mead will have to be reduced, it seems certain that the Imperial Valley will take the main hit. Cohen says it will have to contribute 60 percent of the reduction, or 120,000 acre-feet a year. There things stand. Will the Salton Sea survive? Cohen believes that a much smaller sea with carefully managed wetlands set aside for birdlife could be viable with the much reduced inflow. Nobody really knows. Anyway, it wouldn't be the Salton Sea anymore.

III

When the rivers run dry . . .

the wet places die

8

MEKONG

Feel the Pulse

THE SWIFT BOATS and the GIs are long gone. The Mekong Delta is peaceful now. Vietnam veterans returning to the scene of some of the worst battles of the Indochina Wars stop off at Vinh Long, which has become a holiday resort. The only ambushes they face will be from traders at the nearby Can Tho floating market. But the peace and prosperity of modern-day Vietnam is the source of a growing new concern—the fate of one of the world's last great wild rivers.

Almost half a century of wars in Southeast Asia kept engineers away from the Mekong. Their plans for giant hydroelectric dams on the world's twelfth-longest river, first drawn up in the 1950s, gathered dust. Until now. For in the twenty-first century, dam builders have moved in on the Mekong all the way from the delta to its upstream gorges in China. Tens of millions of people along the river fear that, in consequence, the river's days as the most productive fishery in Asia may be numbered. That, with the dams impeding the river's flow and halting fish migrations, the Mekong is about to lose its bounty.[1]

The main channel of the river is suddenly peppered with hydroelectric-dam construction sites. In Laos, Malaysian engineers are at work on the Don Sahong Dam near the border with Cambodia. Next up will be the Xayaburi Dam, whose hydroelectric power will be for sale to neighboring Thailand; and the Pak Beng Dam, which got the go-ahead in late 2016 despite growing concerns about its impact on fisheries.[2] Also on the drawing board in Laos are dams at Luang Prabang, Pak Lay, Sanakham, Pak Chom, Ban Koum, and Lat Sua. Cambodia has plans for Stung Treng and Sambor, one of the last habitats of the Irrawaddy dolphin.

Way upstream, Chinese engineers, who call the river the Lancang, have been busy for more than two decades constructing a cascade of dams in the steep gorges of the southern Chinese province of Yunnan. After the smaller Manwan and Dachaoshan Dams, completed in 1993 and 2003 respectively, they

Mekong Delta

finished the giant Xiaowan Dam in 2010. At 958 feet high, it is as high as the Eiffel Tower. Its reservoir stretches upstream for 105 miles and feeds turbines that help keep the lights on as far away as Shanghai, 1,200 miles to the east. In 2014, China completed the Nuozhadu Dam, which is slightly less high at 858 feet but has an even larger reservoir.

China plans a total of eight giant dams that will turn the Mekong into southern China's new electrical powerhouse, far exceeding the power produced at Three Gorges on the Yangtze. The dams will be able to store half the entire flow of the Mekong as it leaves China. In future, the upper river's annual monsoon flood that once rushed downstream to Myanmar (formerly Burma), Thailand, Laos, Cambodia, and Vietnam will instead be released gradually, according to China's demands for electricity.

The river's huge flood pulse, one of the natural wonders of the world and the basis of its fecundity, has already been reduced and in future will be largely lost. According to the UN Environment Programme, China's dams are "the single greatest threat" to the future of the river and its ecosystems.[3] "China is acting at the height of irresponsibility. . . . Its dams could sound the death knell for fisheries which provide food for over 60 million people," said Aviva Imhof, former campaigns director at the International Rivers Network.[4]

The Mekong has long been known as the "sweet serpent" of Southeast Asia. It winds for 2,800 miles out of the ice fields of eastern Tibet and through a long series of deep gorges in the mountains of southern China before tumbling down rapids to flood the rainforests of Laos and Cambodia and sliding into the sea through its delta in Vietnam. The Mekong is far from being the world's largest river. Its average annual discharge of 380 million acre-feet makes it fourteenth in the riverine pecking order. But during the summer monsoon, it contains up to fifty times more water than in the region's long dry season. Then it has the third-largest flow of any river in the world, exceeded only by the Amazon and the Brahmaputra.[5]

That variability is the reason for its huge natural wealth. The river has sustained the world's second-largest inland fishery, a mainstay of the region's economy for millennia. It has made the Cambodians, who are among the world's poorest people, among the best fed. This cornucopia has been sustained in particular by one remarkable feature born of the intensity of that flood—the river that runs backwards.

That river is the Tonle Sap, a tributary of the Mekong in Cambodia. It is the beating heart of the Mekong River system. For seven months of the year, the Tonle Sap flows from a lake in the center of Cambodia to join the main river in front of the royal palace in Cambodia's capital city, Phnom Penh. Each June, however, that flow halts. For five months—until November—the river reverses. This happens because the summer floods on the main stem of the Mekong have so increased the Mekong's flow that the river's main channel cannot contain them. The water has to go somewhere, and so it backs up into the Tonle Sap.

For those five months each year, the tributary flows back upstream for some 125 miles into its lake, which expands dramatically, flooding surrounding forests. At the height of the monsoon season, this reverse flow into the Great Lake in the heart of Cambodia swallows a fifth of the Mekong's raging waters, making the tiny Tonle Sap for a while one of the world's biggest rivers, albeit flowing backwards. During this flood, the Great Lake is the largest body of freshwater in Southeast Asia. Its submerged forests become the nursery for the Mekong fishery. In the silty water among the tree roots, billions of fish fry grow into fat adults. Each November, as the Mekong flood abates, the Tonle Sap turns, the lake empties, and the fish swim out, making their way into the main river.[6]

To mark this annual turning of the river, Cambodia has a huge water festival in Phnom Penh. It has been held almost every year since the twelfth century. I watched in 2003 as the diminutive king and queen of Cambodia appeared on the balcony of the royal palace, overlooking the Tonle Sap River where it enters the Mekong. Like fairy-tale monarchs, Norodom Sihanouk and his queen, Monineath, serenaded their subjects—upwards of a million of them—gathered on the promenade of the beautiful old colonial capital. The high spot of the festivities, at the full moon, was one of the world's oldest boat-races. As I watched, around four hundred large canoes, each decorated with paintings of water serpents, rushed down the river to a finish line right in front of the royal palace. Some boats contained seventy frenetic oarsmen, all standing in a line, pounding the water. Altogether, thousands took part in the races, while a million more Cambodians, from across one of the poorest and least urbanized nations on earth, flooded into the city for a weekend of eating and camping on the riverside.

After the Tonle Sap turns to flow the "right way" into the Mekong, the fish from the Great Lake migrate back into the main river and then for hundreds of miles up and down the Mekong—filling the nets that feed tens of millions of people. It is estimated that two-thirds of the fish in the Mekong begin their life as small fry in the Tonle Sap. The most extraordinary product of this fishery is the giant catfish. The protected species grows up to 10 feet long and can weigh a third of a ton. Its numbers are declining, but it still lurks in the flooded forest and in huge hollowed-out pools on the bed of the Mekong, some of them 400 feet deep. The giant catfish occasionally turn up in the nets that fishermen put across the Tonle Sap. During my visit, fish biologists in Phnom Penh were on twenty-four-hour alert to rush to nets if one was caught.[7]

Once the boat festival was over, I took to the water: up the Tonle Sap, to the Great Lake and into the flooded forest. It is a topsy-turvy world. Giant trees poke from the water. Tens of thousands of people live in floating villages that cruise slowly across the lake. French adventurer and author Pierre Loti described Cambodia as the land "where fish grow on trees," and that, more or less literally, is what most of them do. The flooded forest is one of the most productive natural fish-nursery grounds in the world. Each summer billions of fry are swept here on the monsoon surge to feed on the floating vegetation. Then, as the forest drains each autumn, the fattened fish leave. The local fishermen are ready.[8]

Much of the flooded forest is in private hands, divided into around five hundred fishing lots that are auctioned by the government. Many are managed by migrants from Vietnam, who are regarded as the master fishers. The biggest lots can cover hundreds of square miles and bring in $2 million a year to their owners. From October to March, virtually the entire lake is ringed by bamboo "fences" erected by the lot managers to trap fish as they swim out of the forest. The forest is also a magnet for thousands of Cambodia's poorest people, who take their nets into the open-access areas outside the private lots. Many are landless refugees from the great expulsions by the Pol Pot regime in the 1970s. They depend for survival on the free resources of the lake.

I clambered aboard a floating wooden platform that was the office of Sreng Sokhak, a fishing-lot owner with twenty employees. "I've been fishing on this river for twenty years. We used to catch a hundred tons a year. We get nothing like that now," he said. He complained about his neighbors, and about unlicensed fishermen who come in with batteries and electrodes and send an electric charge through the water to electrocute the fish. They were destroying the fishery, he said. Many fisheries scientists I spoke to, however, argued that a much greater danger would be the loss of the forest and waters that make the system so productive. They said that ultimately the fishermen are the true friends of the fish, because they prevent farmers from felling the flooded forest and converting it to rice paddies.

Whatever the truth in this debate, the fecundity of the ecosystem is extraordinary. Gangs of villagers collect the eggs and chicks of cormorants, pelicans, storks, and ibises. Others hunt squirrels and rats, pythons and cobras, turtles and lizards. Up to a million water snakes are netted each year. Traders come to the lake from Thailand, Vietnam, and even China to buy crocodiles, fish, snakes, and monkeys. A Vietnamese monkey farm was buying macaques to breed for Western laboratories. Crocodiles have been hunted to extinction in the lake and forest, but the floating villages now farm them. Stepping unawares out of my boat onto a jetty in Kaoh Chiveang floating village, in the northwest part of the lake, I was surprised by a rush of water beneath my feet. Looking down, I saw between the wooden slats a score or so of young crocodiles, each about 3 feet long, in a cage suspended in the water.

Kaoh Chiveang was an upmarket suburb on the lake. The floating homes were immaculate, with flower and vegetable gardens at their doors, electricity generators hooked up to long lines of car batteries, and TV aerials on every palm-frond roof. The waterways were lined with stores, karaoke bars, filling stations, and elaborate cranes for lifting boats out of the water for repairs. I soon found out the source of the wealth. The village of some three thousand people had at least as many crocodiles. Most were fed on water snakes harvested from the flooded forest all around. A fully grown reptile fetched about a thousand dollars. Every household with the capital to spare had a crocodile cage. The open water might now be empty of the beasts, but the cages probably contained more than ever lived in the lake.

The Mekong has for decades now been a poignant reminder of how the world's rivers used to be before the dam builders got to work. Two-thirds of our rivers, including most of the largest, have dams holding back the natural flood-pulses on their main channels. The Mekong has been the biggest exception. Nets dipped into the Mekong harvest over 2 percent of the entire world catch of wild fish from both rivers and the sea. There is scarcely a mile of riverbank along the Mekong where nobody is taking fish. Yet the river has kept providing. Some sixty million people eat or draw their income directly from the river. The development agency Oxfam says that Cambodia's river fisheries "make a bigger contribution to economic well-being and food security than in any other country."

The question is: for how much longer? The precise effect of the dams will depend on how they are operated. But according to blueprints seen by Western hydrologists, the Chinese dams alone will cut flood-season flow on the lower Mekong by a quarter—enough to reduce the flood pulse in Phnom Penh by at least a third. Chinese officials say this is good news for downstream because it will reduce flood risks and maintain dry-season flows.[9] Hydrologists argue that the loss of the flood pulse is of much greater significance, though they are divided about whether this taming of the river flood will be enough to end the reverse flow of the Tonle Sap. Matti Kummu, a hydrologist now at Aalto University in Finland, calculates that the dams may only delay the Tonle Sap's reversal by about a month. That would still be enough to reduce the area of flooded forest by more than 2 million acres, he said. An equal danger might be to curtail the pronounced dry season on the river. Without that, much of the forest would die, he warned.[10]

Perhaps just as important for the fishery as the changing water flow would be the loss of the river's load of silt.[11] Here the Chinese dams are of even greater significance. Around one-half of the 160 million tons of silt that once came down the river each year began its journey in China. Much of it ended up in the flooded forest around the Great Lake, fertilizing the vegetation and feeding the growth of fish. But according to Kummu, the Chinese dams are already capturing more than half this silt flow. "If the whole cascade of dams is built, it would trap some 94 to 98 percent of the sediment load coming from China," he said. That would become a catastrophe for the fertility of the Mekong flood, for the flooded forest, and for the entire ecological infrastructure on which much of Southeast Asian rural life is built.

The construction of China's dams on the upper Mekong has been a political as well as an ecological travesty. China began blocking the river without any prior consultation with its neighbors. In 1995, Vietnam, Cambodia, Laos, and Thailand—the four downstream nations on the river—formed the Mekong River Commission as a forum to consult on the river's future. China, along with its ally Myanmar, never joined. China still refuses to. It has never even discussed its dam plans in advance with the commission.

Sometimes it has sought to mollify its neighbors during drought years. In 2015, China and the five downstream states set up the Lancang-Mekong Cooperation Mechanism.[12] The following year, China made extra water releases from its dams in response to Vietnamese pleas about low flows in the river delta, which had resulted in a massive influx of salt from the ocean. But the goodwill gesture did not come before 350,000 acres of delta had been damaged.[13] China has also refused to concede that downstream countries have any right to a say in how it runs its dams. Those nations have to go cap in hand. It is a vivid example of China's continuing disregard for its neighbors. Many other parts of the world suffer from similar bully-boy tactics from upstream nations on cross-border rivers. Take the story of Ethiopia's treatment of Kenya, the subject of the next chapter.

9

SEAS OF DEATH

EUROPEAN EXPLORERS FIRST SAW the vast expanse of shining turquoise waters in the desert sands of East Africa in the 1880s. They called it the Jade Sea. Today the 150-mile long lake in northern Kenya, close to the border with Ethiopia, is known as Lake Turkana, after the tribe living on its western shore. Whatever you call it, this hydrological jewel is the largest desert lake in the world. Half a million desert people are dependent on its waters. But Lake Turkana is shriveling fast and could be all but gone within twenty years.

The problem for the Kenyan lake lies over the border in Ethiopia. The source of almost all the water that flows into Lake Turkana, and the reason the lake does not shrivel in the desert sun, is Ethiopia's river Omo. Its water enters the lake almost exactly on the remote international border between the two countries. In early 2015, however, the water stopped flowing when Ethiopia blocked the river way upstream, having completed Africa's tallest dam. The plan is to capture most of the river's flow, passing the water through the dam's turbines to generate hydroelectricity and then irrigating plantations of sugar and other thirsty crops that are being created in the remote bush.

In the first two years after the dam was closed, the water level in Lake Turkana fell by 5 feet. It is drying out first in the shallow north, where reed beds and woodlands are dying and a bay where most of its abundant fish grew is left high and dry. Kenya has barely raised a whimper about the plunder of these waters. Independent researchers warn, however, that a hydrological, ecological, and humanitarian catastrophe awaits. The water level in the lake depends on the balance between inflow from the Omo and evaporation from the lake's surface. The unrelenting Kenyan sun takes 8 feet a year from the lake, and that must be replaced or the lake will die. Sean Avery, a Nairobi-based consultant with over thirty years' experience working with dam builders in Africa, predicts that within a couple of decades, the lake is likely to lose at least half of its volume and could eventually be reduced to two small salty pools. Five national parks in and around the lake, where hippos wallow and crocodiles feast, will lose most of their wildlife. The ecosystems that sustain the precarious existence of the local people will shrivel and die.[1]

The Ethiopian government is unmoved by such warnings. It says that it has the right to harness the waters of the Omo. The 500-mile-long river flows out of the country's central highlands, which are as wet as Scotland. That water is badly needed in other parts of the country that are dangerously dry. In the 1970s and 1980s, Ethiopia suffered catastrophic droughts that left hundreds of thousands of people dead. That cannot happen again, the government says. Preventing another disaster in a country that now has almost one hundred million inhabitants requires the construction, in a steep canyon 400 miles north of Lake Turkana, of the 797-feet tall Gibe III dam. Gibe III is the latest and much the biggest in a series of five planned on the river to capture and distribute the Omo's flow though hydroelectric turbines and irrigation ditches. Its reservoir will eventually be 100 miles long.

When the Gibe III dam was first proposed more than a decade ago, its designers promised that it would only be used for generating hydroelectricity. Water would be held in the reservoir to pour through the turbines when Ethiopia needed the power, but most of it would eventually make its way across the border and into Lake Turkana. Then the plans changed. In 2011, with construction already well under way, the Ethiopian government suddenly announced that the dam's control of river flow would also allow it to divert the water that had passed through the turbines to irrigation projects on its side of the border. Altogether, the government has earmarked almost 750,000 acres of the lower Omo valley for commercial agriculture. First up—and already under construction in 2017—was the 250,000-acre Kuraz sugar plantation.

Thanks to such projects, the traditional cattle-grazing inhabitants of the land on the Ethiopian side of the border are as angry as their compatriots on the Kenyan side. For while the Kenyans will lose their lake and its fisheries, the pastoralist Ethiopians will see their unfenced landscape of bush, woodland, and open cattle pasture fenced and tilled. Many people will be forcibly resettled. The Kuraz plantation is on land occupied by the Daasanach people. Wildlife will also be expelled.

The Ethiopian government has never published an environmental assessment of the impact of the farms and their water abstractions. But the hydrology is inescapable. Avery estimates that—depending on the pace of development, efficiency of the irrigation, and how much water returns to the river from field drainage—the Kuraz sugar project alone could take 30 percent of the river's flow. The other planned farms could increase the take to 50 percent or more.

The lake would shrink until it finds a new stable state where evaporation from a shrinking surface area is reduced to match the diminished inflow. Avery predicts that this would see the lake's volume halved and the water level lowered by 65 feet. At that point, all that would be left would be two small and very salty lakes on the desert floor.

Falling lake levels are not the only threat. From the moment the dam closes, the seasonal cycles of river flow, driven by monsoon rainfall in the highlands, will be replaced by fluctuations determined by Ethiopia's electricity demands. Till now, peak flow on the river has been eight times the lowest monthly flow. The dam's designers intend that the peak flow should be just twice the slowest flow. Ecologists say that, as on the Mekong in Cambodia, the annual flood-pulse is vital for the ecological health of woodlands, wet pastures, and the lake's ecosystems. Jeppe Kolding, a fish biologist from the University of Bergen in Norway, says the pulse flushes into the lake waters rich in sediment, organic matter, and nutrients. These goodies stimulate fish breeding and growth. Without it, even a diminished lake would be barren.

Until recently during a strong pulse, up to 18,000 tons of tilapia were caught annually in the shallow Ferguson's Gulf on the lake's western shore. The Turkana people sell smoked fish at beach markets and ship more across Kenya and as far as the Congo. Or they did. In the first two years after Gibe III's gates closed, the five-foot drop in lake levels caused the shoreline to retreat by more than a mile. Fish catches collapsed. A 10-foot drop will kill it altogether. Other fisheries could emerge in new shallows. But without the flood pulse, they are unlikely to be nearly as productive, says Kolding.

In response to such criticisms, the Ethiopian government has promised to release an "ecological" flood from Gibe III. It would send downstream 0.8 acre-feet per second for ten days each September. That could be a partial lifeline for the lake, says Avery. However, in the first flood season after the dam closed, the promised release did not happen, and meager flows in 2016 did nothing to refill the lake.[2]

The Ethiopian government is bent on modernizing this backward corner of its country. Besides the dam, it is investing in new roads, new clinics, programs to fight the tsetse fly, and a new airport. Cattle herders such as the Mursi—the last tribe in Africa whose women wear large plates through their lips—have been promised regular jobs on the farms. In truth, they don't have much choice in the matter.[3]

The real benefits of the government's plans for Gibe III will go to the nation as a whole. The massive sugar plantation is part of a plan to make Ethiopia one of the world's top ten sugar exporters. The power plant will double the electricity-generating capability of a country in which only a third of people have access to mains power. Ethiopia is rightly determined not to repeat its nightmare of a generation ago, when it suffered so badly from drought. The tragedy is that its way of doing that will dispossess many of its own people, while creating more droughts, and spreading more deserts, over the border in northern Kenya. The Jade Sea looks set to lose its sheen forever.

The lakes of the world are in deep trouble. It is not just the most famous ones, like Lake Chad and Lake Turkana. Hundreds of others are disappearing. They are either being pumped out directly or starved of water by abstractions from the rivers that feed them. In the Andean highlands of Bolivia, the country's second-largest lake, Lake Poopó, which once covered 1,000 square miles, has turned into a salt flat as mining companies divert water. Farmers are emptying Mexico's largest water body, Lake Chapala, which has been the main source of water for its third-largest city, Guadalajara. The Salton Sea, after thriving for a century, now seems doomed. Lakes in central Turkey halved their surface area between 2003 and 2010. Irrigators have entirely emptied Lake Aksehir, and the country's largest freshwater lake, Lake Beysehir, could go the same way by 2040.[4] Iran is suffering especially badly. By 2016, its irrigators had taken 80 percent of the water from Lake Urmia, which was until recently the world's second-largest salt lake. Thousands of people have left the area.[5]

Most lakes form part of wider wetlands that are under at least as much pressure from upstream water abstractions. In the earlier discussions of Lake Chad and the African Sahel, we saw the chaos that can result. Such losses are tragic partly because our modern world is blind to what is going on. Ecologists line up to defend rainforests, and governments throw money at their protection. But wetlands? Even their various names—bog, marsh, swamp—can be understood as pejorative. When Donald Trump went to Washington, DC, his chosen metaphor for cleansing the political establishment was to "drain the swamp," as if nobody could disagree that that was a good idea. How tragic.

From the peat bogs of Scotland to the backwaters of the Mississippi delta, from the lagoons of Venice to the flooded forests of Cambodia, and from the frozen tundra of Siberia to the salt lakes of the Australian outback, wetlands are an in-between world. Sometimes wet and sometimes dry, sometimes land and sometimes water, sometimes saline and sometimes fresh, they change their character with the seasons. Their wealth is tied up in hard-to-measure intangibles such as flood pulses and the fertile silts suspended in their waters. They may stay virtually dry for several years before being replenished by violent floods. Their wildlife is similarly transitory, with migrating birds, fish, and even mammals coming and going. Like rainforests, they are generally not owned by anyone. Their rippling waters, shifting land, and migrating wildlife are a common resource available to anyone willing to brave the elements to get them.[6]

These are all unhelpful attributes in the modern world, where certainty in nature is valued more than fluidity, where floods are a "bad thing," where private property is the universal currency, and where resources constantly on the move are hard to own. In every sense, the wealth of wetlands can slip through your fingers. The modern world wants to enclose, privatize, and tame them, much as the water in flowing rivers is captured by dams. The owners of wetlands, despairing of exploiting the moving feast, often prefer to drain them, fence in the land, and start again. Engineers have long regarded any water draining into wetlands as somehow wasted and ripe for diversion to other human uses. Meanwhile, the wetlands themselves are seen as ripe for draining for agriculture, whether the English Fens, the Dutch polders, or the world's great river deltas or desert oases such as Jordan's Azraq wetland, which has lost 90 percent of its flooded area since the 1970s.

Next to rainforests, wetlands are the planet's most productive ecosystems—for humans as well as for nature. Yet they are disappearing just as fast and with far less fuss: drained, diked, canalized, concreted over, turned into shrimp farms and rice paddies, dammed, dredged, and filled with solid waste. The data are lacking, but recent estimates assembled by Max Finlayson of Charles Sturt University, in Bathurst, Australia, suggest that between 64 and 71 percent of them have vanished since 1900.[7] Some have been replaced by artificial wetlands such as rice paddies and reservoirs. But, says Finlayson, "this gain does not offset the loss of natural wetlands."

The rich ecosystems of wetlands are why fully functioning wetlands become refuges for communities in times of strife. When all else fails, they are where the poor and vulnerable retreat to. We are seeing that today, for instance, in the Sudd wetland on the Nile in South Sudan. Decades of conflict in the region have prevented plans to dig a canal so the Nile can bypass the wetland and deliver more water downstream. Today, the Sudd is a sanctuary for those affected by the civil war in South Sudan. The UN estimated in 2017 that some one hundred thousand people had fled their homes on dry land and taken refuge on islands in the 14-million-acre Sudd. Perhaps another million people live off the bounty of its ecosystem through hunting and fishing.[8]

As one international assessment concluded: "The social benefits of retaining wetlands, arising from sustainable hunting, angling, and trapping, greatly exceeded the agricultural gains" if the wetlands were drained. And yet they are still drained, because the social benefits often accrue to society in general, while the agricultural gains can be captured entirely by the owner of the land. Lead author of the assessment Andrew Balmford of Cambridge University concluded that "our relentless conversion and degradation of remaining natural habitats—including wetlands—is eroding overall human welfare for short-term private gain." As scarce natural resources become rarer, the social benefit from protecting them grows rather than diminishes. Preserving what remains "makes overwhelming economic as well as moral sense."[9]

Let's look at the travails of one such wetland. One you have probably never heard of. For thousands of years, under a dozen different empires, the Hamoun wetland on the border between modern Iran and Afghanistan has been a refuge and a source of food and water in the desert on the Silk Road to China. Its inhabitants, the Sistani people, were generally left to their own devices—until a lethal combination of drought, American engineering, and Taliban fanaticism turned their oasis to dust at the start of the new millennium.

The Hamoun wetland covers 1,500 square miles. Even in bad years, half that area would be wet. Leopards lurked in the marshes, and carp and otters swam in the lakes. The wetland was a mecca for flamingos, ducks, and pelicans migrating from Siberia to the Indian Ocean—and for the Sistani, who engaged in fishing, hunting, farming, and punting across the lakes in traditional flat-bottomed reed boats called *tutans*. They herded their animals across the rich pastures around the lakes and tapped the waters to irrigate wheat and barley, grapes and melons, even a little cotton and sugarcane. As recently as the mid-1990s, they pulled thousands of tons of fish from the lakes each year.

The water for this seemingly idyllic existence came down the Helmand River from the snowfields of the Hindu Kush in the east of Afghanistan. It cascaded through the wetland's three lakes, the first in Afghanistan and the other two in Iran, before eventually evaporating. Unusually in such an environment, each lake contained freshwater, because occasional floods down the river overflowed the lakes and flushed the salt into flatlands to the south.

That changed in 1998 when, for the first time in recorded history, all three lakes dried out. The reed beds vanished, the marshes were replaced with salt flats, and the wildlife largely disappeared. Hundreds of villages on either side of the border were overwhelmed by shifting sand dunes and ravaged by summer dust storms. Old irrigation channels around the lakes disappeared beneath the sand. The drought continued. By 2002, an estimated three hundred thousand refugees were crowded into camps.

It was a major environmental catastrophe. You might not have read about it, however, because the wetland was on a remote border between a pariah state then run by warlords and one belonging to what President George W. Bush in 2002 referred to as the "axis of evil." Only satellite images of a new desert and a few voices from refugee camps revealed what had happened.

The origins of the tragedy went back to the repeated efforts by Afghans and their Western friends over more than a century to modernize agriculture by irrigating fields upstream along the River Helmand. From the start, the risks to the Hamoun wetland were understood. As early as 1926, under a British-brokered deal, Afghanistan promised to leave half the river's water for the Iranians and the lakes. The deal was renewed in 1973 and remains in force to this day. Sadly, it has rarely been honored. The Iranians claim that Afghanistan has regularly taken more than its share. Especially after its ability to capture the river's flow was enhanced by United States engineers, who in the 1970s began digging dams and irrigation channels and constructing the 300-foot-high Kajaki Dam. Once the Taliban seized control of Afghanistan in the early 1990s, the water coming downstream all but gave out, setting the scene for the crisis of 1998.[10]

With unpleasant irony, some of the trucks ferrying water to the refugee camps during the crisis filled up at the same distant reservoirs that had emptied the river in the first place.

In subsequent years, some water fitfully returned to the wetland. How much depended both on rainfall upstream and on the security situation. The latter determined both whether farmers irrigated their fields and whether the

Kajaki Dam could be harnessed for electricity. But the wetland has never fully recovered, and locals deprived of their livelihoods are turning to smuggling or migrating.[11] In 2017, Iran's reformist president Hassan Rouhani promised that reviving the Hamoun was a vital part of his efforts to tackle the country's ongoing water crisis. Things are largely out of his hands, however. The previous year, Afghanistan had announced plans to install new turbines at the Kajaki Dam, and it had begun construction of a new dam, just before the border with Iran. It is hard not to conclude that the wetland is doomed.[12]

10

ENGLAND

Chalk Streams in Peril

FOR A WATER FEATURE IN THEIR GARDEN, most people make do with a fish-pond. The more ostentatious might run to a decorative fountain. Not Victoria Harrison. Beneath the floor of her dining room in the rolling hills of the English county of Hampshire flows one of England's clearest, purest and—especially in the heat of summer—coolest streams.

Her home is an ancient mill in the picture-postcard village of Itchen Stoke and the river is the Itchen, which rises nearby in a shady pond amid the downland north of Winchester, the early capital of England. We headed into her garden, crossing a rickety wooden footbridge that led to wet meadows, gazing down at wild trout and grayling loitering among dark mats of weeds, looking for mayflies hatching among the gravel beds in the shallow stream. Dragonflies hovered above. "This is so wonderful, we hate to keep it to ourselves," Harrison said. "We had six school groups here last term."

The Itchen is probably the best example of one of the most remarkable—yet least remarked—features of the English countryside. Chalk streams bubble up out of the hills and ripple over extensive gravel beds as they meander slowly toward the sea. They are typically shallow and crystal-clear, their alkaline waters immaculately clean thanks to the constant purifying and filtering of the chalk.

Geographers say there are only 210 true chalk streams in the world, and 160 of them are in England. Their distinctive gravel beds are an irreplaceable relic of past landscapes. They were laid down ten thousand years ago by the gushing floods as ice sheets retreated from England at the close of the last ice age. You can find them draining chalk hills from Yorkshire in the north, through the Chilterns and North and South Downs to the Piddle in Dorset in the southwest. The Hampshire Itchen, the nearby Test, the Lambourn in Berkshire, the Wensum in Norfolk, the Darent and Stour in Kent, the Frome in Somerset, and many more are, ecologically speaking, England's equivalent of rainforests. They are its one unique contribution to global ecology. But, the English being the English, hardly anyone ever makes a fuss about them.[1]

Chalk streams are center stage in many of the best-loved images of perfect rural England. The Kennett, a tributary of the river Thames, flows past Wiltshire's Silbury Hill, the largest artificial mound in Europe and as old as the Egyptian pyramids, before meandering on to Avebury's giant Neolithic stone circle. Poet laureate Sir John Betjeman wrote of the Kennett: "When trout waved lazy in the clear chalk streams, Glory was in me." Painter John Constable famously portrayed the Hampshire Avon and its water meadows in front of the spire of Salisbury cathedral. Kenneth Grahame's *The Wind in the Willows* is set on a chalk stream.[2]

Anglers love chalk streams. Fly-fishing was invented to catch trout on the Itchen. Ecologists also love them for their rare species, including the southern damselfly, one of Europe's most endangered insects, as well as kingfishers, the white-clawed crayfish, and plants like the river water-crowfoot. As Harrison waded through the water meadow in her Wellington boots, a startled heron took off from the rushes. Back on the river, her black Labrador rapidly retreated from the water when a swan stopped nibbling the weeds to hiss at it. A pair of otters lived here.

The wet meadows of chalk streams don't just serve ecology. During torrential rains the previous winter, when the people of central Winchester were bringing out sandbags to protect their homes from the Itchen's rising waters, upstream meadows such as that at Itchen Mill helped catch water and slow the flow into the city, reducing the flood peaks.

You would think the English would long since have decided to take proper care of these treasures. The Itchen has, in theory, copper-bottomed protection. It is listed as a special area of conservation. But, says Rose O'Neill, freshwater campaigner for the World Wildlife Fund, threats to their future purity and ecological uniqueness come from farmers and from the demands of water-supply companies. Existing laws to protect them are flawed. The main enforcement agency, the Environment Agency, is hamstrung by political pressure to ensure water supplies for industry and homes.

Only a couple of miles upstream of Harrison's eighteenth-century mill, the waters of the Itchen are diverted to percolate through some of the country's largest watercress farms. Together the farms abstract almost 25 million gallons of water a day. On the face of it, that's fine. Watercress grows wild in the local chalk streams, and the farmed beds look rather like industrialized versions of chalk streams, their shallow pools lined with gravel and constantly irrigated with sparkling river water. Afterwards, most of the water runs back into the river.

The trouble is that the farms are fertilized with nitrogen and phosphorus that change the stream's chemistry and encourage the growth of algae. Worse, says Harrison, the watercress companies have grown so big that they now bring salad vegetables from all over Europe to wash in the stream waters prior to packing. "These huge trucks come here, driving through our lanes, to wash foreign salad in water from our stream," she says with a sudden burst of anger.

A threat that looms even larger, and one that is shared by most of England's chalk streams, is water abstraction for towns. The chalk hills where the streams rise are immense water stores—bigger than any local artificial reservoir. As a result, chalk streams have in the past always kept flowing. Even in the driest English summers, water kept flowing out of the hills. Today that is no longer true. Densely populated Southeast England pumps two-thirds of its water from underground, usually from chalk. Water levels in the hills are falling. The springs are failing and the streams are drying up, killing their unique ecosystems. The Environment Agency has thirty-seven chalk streams on its endangered list because of overabstraction, almost a quarter of the total. Many are shadows of their former selves.[3] WWF says things are even worse, that half of them are at risk.[4]

Just downstream from Harrison's dining room, the local water company takes water from the Itchen to supply more than three hundred thousand homes. The Environment Agency has told the company to reduce those abstractions by putting in a pipe to take water instead from the nearby river Test, another chalk stream. But that would just kill the Test instead. It is almost equally prized by ecologists and fly-fishermen alike. There should be a better way. But what? Can my England—a country that is often keen to tell the rest of the world how to go green—really sit back while it destroys its own "rainforests"?

IV

When the rivers run dry . . .

we mine our children's water

11

INDIA

A Colossal Anarchy

JITBHAI CHOWDHURY FARMS on the edge of a village called Kushkal in the Indian state of Gujarat. He is, by conventional measures, a model farmer. He is ecologically minded too. He uses manure and natural pesticides that he makes on his farm by soaking roadside weeds in water. He grows fruit trees around the edge of his fields and tends his cattle with care. I met him early one morning as the sun burned off the mist over his fields. He was milking his cows into a churn that he took twice daily to a village collection point, from where trucks transported the milk to India's famous Amul dairy cooperative, which is based in the state. It seemed a perfect organic dairy farm.

Probe a little deeper, however, and Chowdhury's very efficiency suggests the madness of the water economics being played out in his state. Milking done, Chowdhury described for me how his farm worked. He had just 5 acres of land—land that would be virtual desert without underground water. He had a small pump that brought to the surface 3,200 gallons of water an hour. It took him sixty-four hours to irrigate his fields—a task that he carries out twenty-four times a year, mostly to grow alfalfa to feed his cows. His farm's main output was 6.5 gallons of milk a day.

I did the math. He must use 4.9 million gallons of water a year to grow the fodder to produce just under 2,400 gallons of milk. That's 2,000 gallons of water used for every gallon of milk produced. That was not bad by local standards. But even so it meant that over the year, he pumped from under his fields twice as much water as fell on those fields in rainfall. No wonder the water table in his village was 500 feet down and falling by about 20 feet a year. What looked at first sight like an extremely efficient local economy, making milk in the desert for a dairy cooperative that traded across India, was in fact hydrological suicide.

"Yes, I'm worried that the water will disappear," Chowdhury told me. "But what can I do? I have to live, and if I don't pump it up, my neighbors will." As we gave him a lift into the village to deliver his milk churn, he added, "I don't want my son to do farming. I want him to get a job in the city." No wonder.

A generation ago, Indians were starving. Their grain stores were empty and the doomsayers were gathering with the vultures. Today the stores are full and the fear of famine has receded, thanks to a green revolution on new high-yielding crop varieties, such as rice, wheat, corn, and alfalfa to feed dairy herds. India, like many of its Asian neighbors, has kept bellies full while its population has doubled. But the key to the success of these crops has been abundant supplies of irrigation water. Many Asian countries, India among them, have achieved agricultural self-sufficiency by emptying their rivers into irrigation canals. Even so, India's green revolution is increasingly being augmented by plundering the country's underground water as well. Tens of millions of Indian farmers now pump from underground reserves to water their fields.[1]

The farmers' self-reliance is born partly of the administrative failure and technical inefficiency of many surface-irrigation projects built by India in the early years of the green revolution. Indian engineers liked building dams and canals more than creating the fiddly works needed to get the expensively collected water to farms and fields or to insure that that water was used efficiently. Their political masters liked grand openings of large structures—often festooned with plaques with their names on them—more than the actual business of feeding their people.

The failure of surface-irrigation projects was not just about administrative indolence, however. India's rivers simply do not contain enough water to sustain the demands being made on them, so millions of farmers have taken things into their own hands. They are hiring drilling rigs and buying cheap electric pumps to mine water that has lain undisturbed beneath their land for millennia. Groundwater withdrawals across India have increased tenfold in half a century.[2] By one estimate, about a fifth of India's electricity is used for pumping water. This hydrological bonanza has kept India fed. And there is plenty of evidence that farmers who take their own underground water use it more efficiently than their neighbors who rely on canal water.[3] But the overpumping is threatening to turn into a full-blown tragedy as water levels underground fall. Eventually, if things go on as they have, there will be no water left.

My visit to Chowdhury's farm, in 2005, came at the height of a worsening water crisis in Gujarat. Afterwards, I discussed the likely consequences with Tushaar Shah, the director of the International Water Management Institute groundwater research station, which is based in the state. Indian farmers were living in a fool's paradise, he said. "It's a colossal anarchy. Nobody knows where the pumps are, or who owns them. There is no way anyone can control what happens to them."[4]

He put the annual abstraction for irrigation in India at about 200 million acre-feet of water a year. That was twice what the rains replaced. Every year, the underground reserves became emptier. The pumps had to work ever harder. The boreholes had to be replaced or sunk to greater depths. Shah estimated that at least a quarter of Indian farmers were mining underground water that nature will not replace. That was two hundred million people facing a waterless future.[5]

Fifty years ago in Gujarat, yoked bullocks could lift water in leather buckets from open wells dug to about 30 feet. Now tube wells were sunk to 1,300 feet. Half the traditional hand-dug wells and millions of tube wells had dried up across western India. In the south, two-thirds of Tamil Nadu's hand-dug wells had failed, and only half as much of the state's land was irrigated as a decade before, he said. There were fifteen thousand abandoned wells around Coimbatore, the state capital. Whole districts in arid states like Gujarat and Tamil Nadu were emptying of people. Many more farmers were joining the millions crowding into urban slums or the gangs of construction workers and laborers traveling the roads of India.

A decade later, I met up again with Shah. He was more optimistic about the situation in Gujarat than he'd been in 2005. On his advice, the state government had at least partially tamed the "colossal anarchy" of groundwater pumping by rationing the electricity used by farmers. It had done this by separating power supplies to farms from those to communities, and then restricted power supplies for farmers to eight hours a day. Water tables were still falling, Shah admitted, but with pumping time limited, the decline was much slower than before.[6] It was a clever response, but whether it will work in the longer run is far from clear. Farmers have been buying bigger pumps, so they can bring more water to the surface during their allotted eight hours of power. The state government has reacted by proposing a smart metering system to directly control how much power the farmers could snaffle.

Electricity rationing could become a model to save underground waters across India, Shah told me. It should have one high-placed advocate. The Gujarat policy was introduced by its then political boss Narendra Modi, who in 2014 became India's prime minister. Three years on, there were few signs of any effective controls on pumping across the country.[7] Much hangs on this. Because in the hydrological anarchy, farmers have been finding ever more profitable—and reckless—ways of extracting the water beneath their land, as I discovered when I traveled south to Tamil Nadu.

Suresh Ponnusami sat back on his large wooden veranda by the road south of the Indian textile town of Tirupur in Tamil Nadu. He was not rich, but for the owner of a 2.5-acre farm, he was doing well. He had a phone and a television. His traditional white Indian robes had been freshly laundered that morning. As we chatted, I noticed an improbably large water reservoir at the side of his house. Then a tanker drew up on the road. The driver hauled a large pipe over the hedge and dropped it into the reservoir. I expected the tanker to start filling the reservoir with water. This was a drought region, after all. Instead, the pump was working in the other direction—the contents of the reservoir began to empty into the tanker.

Ponnusami explained: "I no longer grow crops, I farm water." He had sunk boreholes deep into the rocks beneath his fields and was bringing water to the surface twenty-four hours a day. He used to grow rice, he told me, but apart from cultivating a little fodder for his beloved cattle, he did not bother anymore. He didn't need to. "The tankers come about ten times a day," he said. "I don't have to do anything except keep my reservoir full." He was mining water on an industrial scale for sale. The water table was 1,000 feet underground and falling fast, said Ponnusami, as his wife poured us a cold lime drink. "Before the trade in water got going here a decade ago, you could tap water at about 500 feet, so it's going down by about 50 feet a year." How much water is left? He didn't know, but he figured he had a way to go yet. In Mandaba, the village down the road, they had drilled to 1,500 feet. "But they are starting to run out of water there," he said.

This was a staggering new industry. Around five hundred water tankers drove each day to this small area. Day and night they came, adding their roar to the constant whine of the farmers' Japanese pumps. Ponnusami showed me the books. He sold his water for two hundred rupees, about four dollars, for a single tanker-load. Of that, he spent about fifty rupees on the highly subsidized electricity available to farmers for pumping water. There were also bills for deepening his boreholes each time they ran dry. "But it's a good living, and it's risk-free," he said. No risk of failed crops. Well, at least not until the water runs out—for this crop will one day fail for good.

I asked where all the water went. Who around there wanted five hundred tankers' worth of water every day? He named two companies with whom he had contracts. They both manufactured textile dyes in nearby Tirupur, part of a region known as "the Manchester of India" because of the huge cotton-growing and knitwear industry set up by local entrepreneurs in the 1950s. Town billboards advertised industrial sewing machines and computerized embroidery, peroxide bleaching, dyeing, and the spinning of local silks.

All these factories needed water—the hundreds of backyard dyeing and bleaching factories most of all. They had once got their water from a giant reservoir 60 miles away on Tamil Nadu's biggest river, the Cauvery. Not much water flowed down that river anymore, however. Like many Indian rivers these days, the Cauvery has been reduced to a trickle, and Tamil Nadu's main reservoir is nearly empty most of the year. As a consequence, the factories had taken to buying underground water from local farmers. It is a trade that has been growing all over India. Around the corner from Ponnusami's farm, I came across a drilling rig by the road. Ponnusami's neighbor was probing deeper and deeper into the earth to keep the water flowing for the tankers. The neighbor had a nice two-story house that clearly hadn't been paid for from selling any crops growing in the stony land around it. Sure enough, soon another tanker driver was loading up from a big concrete-lined reservoir, which was itself being filled from two boreholes. The driver, who collected water from several local farms, said he sold the water in town for four hundred rupees—a clear 100 percent profit.

As we talked, another tanker drove up the track to the next farm. That was three tankers pumping from three farms within 300 feet of each other on one morning in the middle of the Indian countryside. The farmer's wife came out to survey the scene. "Every day the water is reducing. We drilled two new boreholes a few weeks ago and one has already failed. But we will keep drilling till we only have enough water for drinking," she said. Surely this is madness, I suggested. Why not get back to farming before the wells finally run dry? "If everybody did that, it would be well and good," she agreed. "But they don't. We are all trying to make as much money as we can before the water runs out. If we stopped pumping just on this farm, it wouldn't affect the outcome."

This madness is a classic case of what environmentalists call "the tragedy of the commons." Everybody chases short-term wealth even at the cost of destroying their long-term collective future. Nobody can afford to miss out on the boom, because they will all share in the eventual bust. It is certainly what has been happening to India's underground water. Some think it is what we are doing to the planet.

12

OVERPUMPING THE WORLD

THE HIGH PLAINS are part of American history. The first white settlers going west transformed the plains from buffalo hunting grounds into rough pasture for cattle—the land of the cowboy. Then came the plow, and pasture became dry prairie. Until the droughts and dust storms of the 1930s blew the soil away. Millions of sharecroppers abandoned the land and trekked on to California. Since then, the arid plains have been transformed once again—through the pumping of water from a giant underground reserve discovered beneath them.

Known as the Ogallala Aquifer after the Sioux Nation that once hunted buffalo across the plains, the underground reservoir stretches beneath most of Nebraska, South Dakota, western Kansas, Oklahoma, and Texas and parts of eastern New Mexico, Colorado, and Wyoming. In the 1930s, just six hundred wells tapped this aquifer. By the late 1970s, there were two hundred thousand wells, supplying 22 million acre-feet of water a year to more than a third of the country's irrigated fields. Many got rich from exploiting the aquifer. Men such as Clarence Gigot, who bought up huge tracts of the Nebraska Sandhills at rock-bottom prices and became the aquifer's biggest pumper. Some just made themselves richer—like oil entrepreneur Armand Hammer, whose Iowa Beef Packers made its fortune in South Dakota.[1]

The aquifer was an enormous United States resource but also a global one. In a good year, the High Plains could produce three-quarters of the wheat traded on the world market, feeding starving Ethiopians and keeping Egyptians fed as the Nile ran dry. The US became the world's biggest exporter of virtual water—but at the expense of draining the Ogallala.

The problem is that much of the aquifer is to all intents and purposes a fossil resource laid down in wetter times. Little water is added from today's scanty rains. Wells have been drying out now for more than thirty years. The first were in Deaf Smith County, in the Texas Panhandle. During the summer of 1970, a well that had been pumping water since 1936 suddenly went dry. Many others have followed. In large parts of the aquifer, two-thirds of the water is gone. All new well-sinking is banned in some places, and fewer than 10 million acre-feet are now pumped annually, less than half the output in the 1970s. Even so, predicts Laurence Smith of the University of California at Los Angeles, "in north Texas the aquifer will run dry in the next 20 years." Fly over the land and you can see the circular marks where rotating sprinklers once kept the soil wet and the fields green—but the soil is now dry and the fields are brown. The sagebrush and buffalo grass are returning. The buffalo may follow.

This should be a wake-up call for the whole country. A third of all the US's irrigation water comes from underground, and some states in the South and West would be more or less literally lost without it. In Arizona, the Southwest Aquifer is virtually the only water source within the state. Overpumping in the West is as widespread as on the High Plains. The combined annual overpumping of the Ogallala, California's Central Valley, and Southwest Aquifers has been 30 million acre-feet, resulting in a cumulative loss of underground water storage of over 800 million acre-feet. Where the US leads, China and the rest of the world follow.

Every year, more than a hundred million Chinese eat food grown with underground water that the rains are not replacing. In the Pakistani province of Punjab, which produces 90 percent of that country's wheat, farmers pump 30 percent more than is recharged, and water tables are plunging by up to 6 feet a year. Tens of millions of farmers across the world are taking advantage of advances in drilling technology that allow them to access water far deeper than they could with their old hand-dug wells. Asia leads the way in this unsustainable groundwater revolution, but countries such as Mexico, Argentina, Brazil, Saudi Arabia, and Morocco are increasingly significant players. Sandra Postel of the Global Water Policy Project, based in New Mexico, calculates that a tenth of the world's food is being grown using underground water that is not being replaced by the rains.[2]

When a river runs dry, it is very visible. Underground water, by contrast, is invisible. Only the farmers know they have to drill deeper and deeper to find it. And few in the corridors of power talk to farmers about a slow-burning disaster that will one day affect hundreds of millions of people. That is why the world's underground water reserves continue to deteriorate rapidly. In 2017, Yoshihide Wada of the International Institute for Applied Systems Analysis warned that overpumping had increased by a quarter between 2000 and 2010.

It had doubled in China. He estimated that the crops most to blame were rice, responsible for 29 percent of the overpumping, followed by wheat at 12 percent and cotton at 11 percent.[3] Across the Middle East and North Africa, overpumping is routine and largely unnoticed. Few countries have much idea how many wells their farmers have, let alone any means to regulate them.[4]

Major cities—among them Beijing and Tianjin, Mexico City and Bangkok—are also growing increasingly reliant on pumping out underground reserves. This cannot go on. The end won't happen everywhere at the same time, of course. Each aquifer has its own countdown to extinction. But as each aquifer dries up, it will undermine the world's ability to feed itself.

It is true that new water is still being found in the bowels of the earth. In recent years, Chinese scientists have reported the discovery of unsuspected water beneath a vast system of sand dunes in the Gobi Desert, which they claim is being replenished from mountains to the south. New estimates suggests that the Guarani Aquifer, which stretches beneath more than 400,000 square miles of Brazil, Argentina, Uruguay, and Paraguay, may contain 40 billion acre-feet of water—as much as flows down the Amazon River in seven years. The aquifer is already supplying fifteen million people with water without any general decline in water tables. Unlike many of the world's great aquifers, it is still being replenished by the rain. Hydrologists now believe it could one day supply as many as two hundred million people. They want to build an aqueduct to take some of its largesse to the world's third-largest city, São Paulo.

It is also fairly clear that there is a huge amount of untapped groundwater in much of sub-Saharan Africa, where abstraction for irrigation is much less than in Asia. The continent is often thought of as short of water. The images of hungry people searching for food in droughts are seared in our memories. But when the International Water Management Institute mapped Africa's groundwater potential, it reckoned that Africa could raise its groundwater irrigation from 5 million acres to 100 million without overpumping.[5]

Maybe so. In practice, however, newly discovered aquifers will often be quickly mined to exhaustion. That seems set to happen to a large underground water reserve west of Lake Turkana in the Kenyan desert that was discovered by remote sensing in 2013.[6] According to a study for UNESCO, the Lotikipi Basin Aquifer contains about 160 million acre-feet of water, roughly the same as sits in Lake Turkana. Politicians were jubilant. "This water opens a door to a more prosperous future for the people of Turkana and the nation as a whole," declared Kenyan environment minister Judi Wakhungu. Not so fast, say hydrologists familiar with the find. Much of the water is more than 300 feet below ground, so pumping it to the surface will be expensive. Moreover, in the dry lands of eastern Kenya, recharge of the aquifer will be very slow. Once emptied it would take two hundred years to refill.

Some say ancient underground water that is not being replaced by the rains should never be pumped, except in dire emergencies. This "fossil water" should be maintained as a precious backup reserve, especially when climate change seems set to make surface sources increasingly unreliable. Others argue that new water reserves will always be found, so we should not worry so much about running out.

That has been the view taken in Libya, the driest nation of its size on earth. When the coastal aquifers that its farmers had long relied on to grow their crops gave out, the country's leader, Muammar Gaddafi, decided in the 1970s to drill into the giant Nubian Sandstone Aquifer beneath the Sahara Desert. The aquifer is the largest liquid freshwater source on earth. It contains around 50 billion acre-feet of water in a series of basins whose only outlet had been oases in the desert. Most of the water got there during an era when the Sahara was a crocodile-infested swamp. That era ended abruptly seven thousand years ago, when the desert formed. Since then, except for a little infiltration from the Tibesti Mountains in northern Chad, this "fossil water" had sat undisturbed. Until geologists discovered it while prospecting for oil in the 1960s.

Gaddafi, who seized power in 1969, decided to spend his oil revenues on bringing the water to the surface. At first he tried to persuade Libyan farmers to move to his boreholes to grow wheat. The remains of Gaddafi's giant desert farms can still be seen from aircraft crossing the desert. The great migration to the desert never happened, however. Libyans preferred their fields along the Mediterranean coast, even though the aquifers that watered them were emptying fast. So the colonel decided instead to send the water to the people. He began constructing a 2,000-mile network of pipes, all of them wide enough to drive an underground train through, that he called the Great Man-Made River.

The plan was to carry 600,000 acre-feet of fossil water each year from well fields around the oases of Sarir and Tazirbu on a nine-day journey across 600 miles of desert so dry that even camel trains didn't cross it until the nineteenth century. There was a huge ceremony on the coast near Sirte, Gaddafi's hometown, in 1991 to greet the first water. Leaders from Africa and the Middle East gathered in the cool desert night to hear the colonel compare his great work to the pyramids of ancient Egypt and watch him unleash the water into a reservoir.

Gaddafi boasted that day that his new river was a revolutionary triumph against the tyrants of the United States and Britain, who had bombed Tripoli just five years before. What he neglected to mention was that the project was being masterminded by the enemy. It was being managed not from Tripoli but from an office on the fourth floor of a top-security high-rise near the train station at Hampton Wick, a leafy suburb of southwestern London. There, through the bombings and kidnappings and sieges and murders that characterized Anglo-Libyan relations in the 1980s, and in apparent contravention of a US embargo on trade with Libya, Brown & Root, a European subsidiary of the American engineering corporation Halliburton, worked with Gaddafi's staff to make the colonel's pipe dream a reality. When I visited those offices, Bashir el-Saleh, the colonel's US-trained head man in Hampton Wick, said, "This is the best investment we have ever made. Hopefully we will be able to depend on it for hundreds of years."[7]

The project, for some years the world's largest civil engineering project, had used some 5 million tons of cement and 25 million tons of aggregate. It cost $27 billion, draining Libyan oil revenues for more than two decades. Conspiracy-minded foreigners sometimes couldn't quite believe what was going on. The *New York Times*, in 1997, ran a story quoting defense analysts who claimed that the pipes were not carrying water at all. They had been built to hide Libyan tanks embarking on some future invasion of the country's neighbors, and to provide Gaddafi with "somewhere to store chemical and biological weapons."[8]

That was nonsense, of course. By the time Gaddafi was overthrown, in 2011, the Great Man-Made River was 70 percent complete and definitely delivering water. It even survived the chaos of the civil war leading to his ouster and death, during which NATO planes, acting in support of rebels, destroyed one of two pipe-making factories.[9] To the surprise of many, despite the breakup of the country into a series of enclaves run by rival militias, the pipeline network largely continued to function. Engineers stayed at their posts. The main problem was power shortages that interrupted pumping, leaving faucets dry in cities such as Tripoli and Benghazi. But when power operated, the desert water continued to reach coastal faucets and irrigation canals. It was one of the few things in the country that continued to work.[10]

Oil-rich states in the Middle East have a growing reputation for spending their money on foolhardy projects to pump underground water from beneath their deserts. Saudi Arabia, like Libya, has virtually no rain and no rivers or surface lakes of any kind. It has spent $10 billion on desalination works to fill its faucets. Powering the works reportedly consumes around a quarter of the country's oil. Then, during the 1980s, the government spent another $40 billion of its oil revenues on sinking pumps into the vast aquifer beneath the Arabian Desert, and marking out 2.5 million acres of land for wheat farms to use the water. Soon, sprinkler arms 300 yards long swept over the sands, distributing water pumped up from as much as half a mile below. Money was no object. The government charged nothing for the water pumped so expensively from beneath the desert, and virtually nothing for the fuel needed to pump it. Then it bought wheat from its farmers at five times the international price—not just the wheat the nation needed but any wheat the farmers sought to produce. Few cared that most of the water that reached farms evaporated in the sun. For every ton of wheat grown, the government supplied 2.5 acre-feet of water—three times the global norm.[11]

To make matters worse, from the mid-1990s, many Arabian farmers converted their wheat fields to alfalfa fields—which used even more water. The alfalfa filled the feedlots for a new national obsession with dairy cattle. High-tech cowsheds sprouted across the desert. Inside them, fine mists of water kept the animals cool in the baking desert sun. Just as wheat once grew in the desert, now Holstein cattle were milked in the land of the camel.

There was undoubtedly a lot of water beneath the Arabian Peninsula. At the start of the 1980s, the country calculated the proven reserves at around 500 million acre-feet, enough to fill Lake Erie. But virtually none of it was being replaced by rainfall. The pumping was so intense that by the turn of the century, hydrologists estimate that only a fifth of that irreplaceable reserve remained. At that point, the Saudi government announced a U-turn. It would end wheat subsidies. Instead, it would import wheat to make Saudi bread. It decided to keep the cowsheds but reduce the animals' water needs by feeding them on foreign fodder. Under the King Abdullah Initiative for Saudi Agricultural Investment Abroad, the sheikhs decided to buy up long-lease farmland in foreign countries. Between 2004 and 2009, Saudi Arabian companies leased some 1 million acres of Sudan to grow wheat and rice, and took more in the Philippines, Cambodia, Pakistan, and elsewhere. Saudi Arabia had at a stroke become the world's leading international land-grabber. But it wasn't really grabbing land; it was grabbing water.[12]

At least Saudi Arabia has options when its underground water reserves falter. It has money to buy food and land on the open market. The people of Gaza have few such options. The densely packed Palestinian enclave, a narrow coastal strip between Israel and Egypt, is running out of drinkable water. Aside from Kuwait, which lives on desalinated seawater, Gaza's million-plus inhabitants have the lowest per capita availability of natural freshwater in the world.

Palestinians pump at least 100,000 acre-feet a year from their aquifer beneath the sands. The recharge rate is about 73,000 acre-feet. Not surprisingly, the water table is plunging. Meanwhile the porous rocks that were once full of freshwater are being infiltrated by sewage from Gaza's towns and refugee camps, and by salty seawater from the Mediterranean. The aquifer's bacteria count and salinity level rise inexorably, making the water mostly impossible to drink and increasingly poisonous for crops.[13]

The crisis is, arguably, partly of the Palestinians' own making. Two-thirds of the water is brought to the surface by Palestinian pumps for irrigation, much of it from unlicensed wells. But they are not the only people with pumps. Israelis farm the Negev Desert right up to the Gaza border. Seen from Gaza, their greenhouses glint in the sun, guzzling up water that could be growing Palestinian crops. The Israelis take water from the aquifer and from the Wadi Gaza, a desert drainage basin that begins on the West Bank south of Jerusalem, then passes through the Negev Desert in Israel and flows through a small coastal wetland in Gaza to the sea. In the past, water from the wadi replenished the aquifer through the wetland. In recent times, however, Israelis have built dams on the wadi (which in the Bible and on Israeli maps is called Besor), so little water reaches Gaza.[14] The wetland has turned into a sewage sump, spewing onto the beaches of Gaza.[15]

Israel and aid agencies deliver some drinking water to Gaza in trucks. Even so, a UN study in 2014 concluded that, without a water rescue plan, Gaza would be uninhabitable by 2020 because there would be no water. There are plans to recycle drainage water from irrigated fields, to catch more of the sparse rains, and to build a desalination plant. Even so, with no political settlement in the region, the chances of delivering the investment for such projects remains low.[16]

13

BANGLADESH

The World's Largest Mass Poisoning

THE CARCINOMA ON Abdul Kasem's hand was deep and gruesome. Doctors first saw it when he visited a clinic in southern Bangladesh the week before. He also took a flask of water from his home well. He wanted checks on both. His medical diagnosis had been instant, but I met him with a health worker who brought the results of the test on Abdul's drinking water. Sure enough, the well from which he had drunk for twenty years contained a lethal 500 parts per billion of arsenic.

We met Kasem outside Barai Kandi, a village of rice growers deep in the delta of the river Ganges. He took us to meet his wife and five children. All six, along with another twelve villagers I met in nearby houses, had warts and other growths characteristic of arsenic poisoning. The health worker gave them prescriptions for vitamins A, C, and E. "It might help, and it's all I can do," he said. The villagers then collected samples from their backyard wells and conducted a simple test for their arsenic levels. The results ranged up to a horrific 1,000 parts per billion. None passed the World Health Organization (WHO) limit of 10 ppb. Before we left, an impromptu village meeting persuaded the owner of the two cleanest wells to let his neighbors take drinking water from him.

Tens of millions of people across Bangladesh and eastern India have for years been drinking well water laced with arsenic at concentrations that are likely to kill them. Often they have done this without knowing it, because there are more than twelve million wells in Bangladesh alone, and tests on their water have been haphazard to say the least. WHO has called it "the largest mass poisoning of a population in history."

The arsenic comes from rocks of the Himalayan Mountains. Over thousands of years, great rivers like the Ganges and the Brahmaputra that rise in the Himalayas have eroded the toxic metal and washed it downstream. It accumulated in the thick muds of the rivers' floodplains and deltas. There it stayed undisturbed, until humans over the past 50 years began to sink bore-holes into the mud and pump up the water through tube wells for drinking. Wells sunk to depths of between 60 and 300 feet brought up the highest concentrations of arsenic dissolved in the water. Catastrophically, that is just the level to which most tube wells have been sunk. At least half of Bangladesh's backyard tube wells, with their simple hand pumps, are thought to be poisoned. Many deliver water with hundreds of times the WHO limit for arsenic in drinking water. The impacts were not immediate. Arsenic is a cumulative poison. It typically takes a decade of drinking arsenic-laced water for the first symptoms to appear. But for more than two decades now, a silent epidemic has been stalking Bangladesh's sixty-eight thousand villages. Tens of thousands of Bangladeshis have already developed skin lesions, cancers, and other symptoms. Many have died.

Almost any tube well could be delivering lethal concentrations. Seemingly small differences in location and depth can make big differences in the arsenic content of the water. And even when wells have been decreed dangerous, villagers often have no alternatives. The biggest scandal is that the wells were mostly installed with aid money from the British and other governments, from charities, and from UN agencies like UNICEF, which sunk the first 900,000 wells. The aim was to improve local health by reducing the heavy death toll from sewage-borne bacteria and diseases spread by contaminated surface waters. In the 1970s, these diseases were killing an estimated 250,000 people a year in Bangladesh. So, many people are alive today as a result of the sinking of the tube wells. But the arsenic epidemic could have been prevented if the risks had been realized in time. Tube wells could have been sunk deeper, to where the water is generally free of the poison. At the least, the water could have been tested before being used for drinking. Neither was done. Two decades after the epidemic was first uncovered, the poisoning continues.

After the crisis first emerged in Bangladesh in the late 1990s, the government launched a mass screening of some five million tube wells. It ran from 2000 to 2003. Afterwards, many of the agencies that funded the initial tube-well program paid to install some two hundred thousand new deeper tube wells and piped water supplies in villages. But the work tailed off after 2006. A study published by WHO in 2012 found that as many people were still at risk from arsenic in their water as had been a decade before. Around 45 percent of water from village tube wells, drunk by twenty million people, still exceeded the WHO guideline level. The authors put the likely death toll at forty-three thousand each year from cancers and cardiovascular and lung diseases caused by the arsenic. That meant that more than one in twenty of recent deaths in the country has been due to arsenic poisoning.[1]

A 2016 Human Rights Watch report found that the program to sink deeper wells to find safe arsenic-free drinking water a decade before had been badly managed. Corrupt local politicians had lobbied for piped water for their own areas, the result being that "the government . . . expend[ed] considerable resources in areas where the risk of arsenic contamination [was] relatively low," the report concluded.[2] Also, many of the new wells were still laced with arsenic because they had been badly sited. "The government acts as if the problem has been mostly solved, but unless the government and Bangladesh's international donors do more, millions of Bangladeshis will die from preventable arsenic-related diseases," the NGO said in a statement.[3]

More and more places turn out to suffer from arsenic poisoning from their water supplies. I first heard about it in the late 1980s from Dipankar Chakraborti, the director of environmental studies at Jadavpur University in Calcutta, who came to London to lecture at a meeting of geologists. He had found a growing epidemic in West Bengal, the Indian section of the Ganges delta. For a while it was assumed to be restricted to that area. When he moved his research across the border into Bangladesh, however, he found the same thing, on an even bigger scale. I helped bring the epidemic to wider attention after a visit in early 1998.[4]

In 2003, Chakraborti suggested that arsenic might also be lurking in underground waters fed from the river Ganges farther upstream. Evidence soon emerged from Bihar, just over the state line from West Bengal. Kuneshwar Ojha, a schoolteacher living in a tiny village close to the Ganges, became concerned after his wife and mother both died of liver cancer. He noticed that other family members had developed skin lesions, which he thought looked like symptoms of arsenic poisoning. He took samples of water from the family tube well to Chakraborti, who confirmed high arsenic concentrations. It quickly emerged that eighteen young people had died from apparently arsenic-related illnesses in the village in the previous five years, and a hundred more were sick with early symptoms. The only fit people were the Dalits, or "untouchables," who were not allowed to drink water from village tube wells because of their low social status.

When Chakraborti surveyed wells across a wide area of Bihar, he found that 40 percent contained arsenic above the safe limit. More than half the adults he examined showed symptoms of arsenic poisoning. His findings were well-publicized, but little was done to prevent the epidemic. A study in 2014 found that half of Bihar's thirty-eight districts were still affected by arsenic in their water supplies.[5] State records showed that cancer rates were much higher in high-arsenic districts such as Bhojpur, Buxar, and Bhagalpur.[6]

In 2005, I first heard rumors of arsenic in the water supply from neighboring Uttar Pradesh, and they were confirmed by scientific studies published five years later.[7] In 2014, heath professionals announced that the metalloid was endemic in water supplies in thirty districts in the state. But there was little evidence of any urgency to tackle the issue.[8]

Is there anything special about the Ganges? Maybe not, says Michael Berg, of the Swiss Federal Institute for Aquatic Science and Technology, who reported a decade ago on high levels of arsenic in well water from the delta of the Red River around Hanoi in Vietnam. He found concentrations up to three hundred times the WHO limit. The first tube wells had been installed seven years before, and they seem to have been sucking arsenic from the river into the aquifers.[9] The delta is home to eleven million people. Scientists have warned that the growing use of tube wells to provide water for booming Hanoi could be creating a health time bomb.[10]

Berg and his colleagues have developed an arsenic risk-mapping system, using geological warning signs. It reveals the two most susceptible landscapes—delta regions with new river sediments, such as those on the Ganges and Red River; and alkaline inland drainage basins. The team tested its model in China and identified two previously unsuspected areas likely to be at risk.[11] Twenty million Chinese people were potentially in danger of arsenic poisoning from drinking well water, with high-risk areas in Xinjiang, Inner Mongolia, Henan, Shandong, and Jiangsu, they concluded. It began to look like a pandemic—one that can only grow as rivers run dry and more and more people resort to underground water for their drinking needs.

V

When the rivers run dry . . .

engineers pour concrete

14

WONDERS OF THE WORLD

CHINA IS BUILT ON SUPERLATIVES. A vast infrastructure for an industrialized country of 1.3 billion people was constructed in next to no time. The Chinese poured more concrete between 2011 and 2013 than the United States did in the entire twentieth century.[1] Some 110 million cubic feet of the stuff went into one structure: the Three Gorges Dam on the river Yangtze. That is more than ten times more than in Uncle Sam's equivalent hydrological icon, the Hoover Dam on the Colorado.

Three Gorges is by some measure the world's most powerful hydroelectric dam, with a generating capacity of 22,000 megawatts. The concrete wall is 1.2 miles wide and barricades the world's third-longest river. Completed in 2003, it created a reservoir 400 miles long that drowned the river's famous gorges, stretching all the way back upstream to the new megacity of Chongqing. It flooded hundreds of cities and towns and inundated farmland, forcing the evacuation of around two million people to higher ground. But already the reservoir is filling with silt brought down from the river's headwaters in Tibet and deposited in the slow-moving waters behind the dam. In little more than a decade, banks up to 200 feet high have formed.

Downstream, the dam has lengthened the season of low flow on the river, during which bankside lakes dry up. The lack of sediment in the river threatens a delta where some fifty million people live, including the inhabitants of the megacity of Shanghai. With little sediment to maintain it, the delta is being eroded by the sea. China is benefiting from the massive amounts of hydroelectricity generated at the dam, but the long-term impacts of Three Gorges are only beginning to be felt.

≋

Once the superlatives were all about the Hoover Dam. It was the world's first superdam, and a marvel of its day. The plug across Boulder Canyon, completed in 1935, was taller than a sixty-story building, and bigger than the Great Pyramid of Egypt. Behind it grew Lake Mead, which could hold more than twice the river's annual flow. As Francis Crowe, the surveyor on the project, later put it, "I was wild to build this dam . . . the biggest dam built by anyone anywhere."[2] This was not the language of technical reports, but it was perhaps the truest reflection of the motivation behind large dam construction on the Colorado and subsequently around the world.

The white concrete of the Hoover Dam became a symbol of Franklin Roosevelt's New Deal public-works projects, and of a broader lust to remake the landscape. It was soon joined by the equally talismanic Grand Coulee Dam on the Columbia River. Famed folk singer Woody Guthrie even wrote a collection of tunes, the Columbia River Songs, for the Columbia River and the Grand Coulee Dam; in it he called dams the newest wonders of the world. They ushered in a postwar world in which dams became symbols of modernism, of economic development, and of mankind's control over nature. The Grand Coulee provided the huge amounts of power needed downstream at Hanford to manufacture plutonium for the first atomic bombs. The Soviet Union wanted its own Hoover Dam; so did Egypt and Japan and China and India. Hoovers sprouted across Latin America; Britain built replicas for its colonies as parting gifts before independence. No nation-state, it seemed, was complete without one.

Dams were also ego-builders for engineers and politicians alike. It is perhaps no surprise that, despite the democratic idealism of the early days of superdams, autocratic, corrupt, and militaristic governments have come to like them best. Marshall Goldman, the analyst of Soviet Russia, identified "an almost Freudian fixation. . . . Nothing seems to satisfy the Soviets as much as building a dam."[3] Under the Fascist leadership of General Franco, Spain built more dams than any nation of comparable size on earth. Today, China has almost 50 percent of the world's large dams, and its construction companies have become the busiest dam-builders around the world, especially in Africa.[4]

The decisions to build have overwhelmingly been "political, benefiting particular politicians or their benefactors rather than solving a problem," says Daniel Beard, former commissioner of the Bureau of Reclamation, the US government agency that built the Hoover Dam and others, more than any other agency in the world. Beard later came out against these projects. When I met him in Japan in the 1990s, he was waving a placard opposing the damming of a major river there. He said he had become hostile to almost all large dams, as had many at his old agency. "In our experience, the actual total costs of completing projects exceeded the original estimate typically by 50 percent," he told me. "And the actual contribution made to the national economy by these dam projects was small in comparison to the alternative uses that could have been made with the public funds they swallowed up. We are now spending billions of dollars to correct the unanticipated impacts such as lost fisheries, salinized soils, and desiccated wetlands." Rivers today are worth more in America as amenities for fishing and tourism than as water for filling reservoirs, Beard said: "We've started tearing down dams."

Was the United States a special case? Far from it. "In my view, we are starting down a similar path throughout the world," he said. "The time when large dam projects are a realistic answer to solving water problems is behind us."[5]

Like Beard, modern environmentalists have come to see large dams as engines of environmental destruction. This is a big shift. For many years, greens in both North America and Europe supported the dam builders. They agreed with India's first prime minister, Jawaharlal Nehru, who called his dams "the new temples of India, where I worship." Rather like wind power today, hydroelectricity was seen as a new, clean, cheap source of electricity. In Europe, concern that dams might interfere with the natural flow of rivers such as the Danube, the Rhine, and the Rhone was tempered by the fear that the alternative was pollution from fossil-fueled power stations. As late as 1996, the head of the World Wildlife Fund (WWF) in Austria told me: "Even though we love the mountains, most environmentalists in Austria still support the construction of dams in their valleys."

A beguiling mantra is that many poor nations, particularly in countries with erratic rainfall, must have dams if they are to grow rich. The World Bank has argued that the amount of water-storage capacity behind large dams on major rivers correlates well with GDP.[6] Kristalina Georgieva, a former environment director at the World Bank, once told me: "Look, countries need infrastructure, and that includes dams. Ethiopians have 13,000 gallons of water storage each; Australians, with a similar climate, have almost 800,000 gallons." Sixty times as much. Americans, she might have added, have double that capacity.

This is sloppy cherry-picking of statistics, replied anti-dam campaigner Paddy McCully. "Zambia and Zimbabwe have around twice the water-storage capacity per person that Australia has, but they remain desperately poor. Paraguay generates nearly ten times more hydroelectric power per capita than Australia but has a per capita GDP just a tenth as big." You might equally make the case that dams bring poverty. Of the seventeen countries that today depend on hydropower for more than 90 percent of their electricity, only Norway is in the top fifty richest nations. However, fourteen are in the bottom hundred: countries such as Burundi, Rwanda, both Congos, Malawi, Afghanistan, and Laos.[7]

Look at the tiny landlocked southern African state of Lesotho. It is one of the most water-rich nations in Africa, with the tallest dam on the continent. That dam, completed in 1986 on the headwaters of the Orange River, has enough reservoir capacity to give each of the country's two million citizens about 400,000 gallons of water a year. Not that they'll benefit from it. Half the people of Lesotho live below the international poverty line, and the nation's main source of income remains remittances from its menfolk working in neighboring South African mines. In early 2004, Lesotho faced famine as parched crops withered in the fields. They could not be watered because almost all the water stored in the mountain kingdom's two giant reservoirs was earmarked for sale to South Africa. Nothing had changed in 2016, when aid agencies warned of impending famine in Lesotho as crops again failed due to water shortages.[8]

So how do we draw the balance between the costs and benefits of big dams, the goods and the bads? Hydroelectric dams do generate a huge amount of power. Around a fifth of the total global electricity comes from turbines on rivers. More than sixty countries depend on them for more than half their power. After China's Three Gorges Dam on the Yangtze, the next biggest, at Itaipu on the Parana River between Brazil and Paraguay, delivers 12,600 megawatts. It supplies São Paulo and Rio de Janeiro, two of the world's megacities.

Most modern dams claim to serve double or triple functions, also supplying water for irrigation projects and city faucets and sometimes flood-prevention capabilities as well. But in practice, one or other function has priority. A hydroelectric reservoir needs to be kept as full as possible, to keep the turbines operating, whereas a flood-prevention dam needs to be kept as empty as possible.

The world's large dams now hold more than 5.5 billion acre-feet of water. Most big river systems, including the twenty largest and the eight with the most biological diversity—the Amazon, Orinoco, Ganges, Brahmaputra, Zambezi, Amur, Yenisei, and Indus—all now have dams on them. On rivers like the Colorado, the Volta in West Africa, and the Nile, dams have reservoirs that can hold two or three times the river's annual flow.

The hydrological, ecological, and social effects of such river engineering have been huge. For many years their status as symbols of modernism insulated them from serious appraisal of the downsides. Even as recently as the 1990s, fewer than half of all proposed dams had an environmental-impact appraisal before construction began. Only since the late 1990s have serious steps been taken internationally to establish whether their benefits outweigh the costs. Even today many social and environmental assessments are barely worth the paper they are printed on. They are heavily biased toward the outcome wanted by the people commissioning them: dam builders. An assessment that recommends the dam not be built is virtually unheard of. The author would never be hired to write another.

In the 1990s, the World Bank, which in the second half of the twentieth century spent an estimated $75 billion on building large dams in ninety-two countries, began to question the wisdom of these projects. Its own internal cost-benefit analyses cataloged huge cost overruns, billion-dollar corruption scandals, poor design and bogus hydrology that left reservoirs empty, turbines never connected to national grids, and irrigation projects that never got built. Bank-financed dams, moreover, had caused the forced resettlement of ten million people. Amid a rising chorus of NGO opposition around the world, the bank pulled out of a high-profile dam on the Narmada River in India. In a quandary over how to proceed, it appointed a World Commission on Dams to assess the successes and failures of large dams and come up with some ground rules for what a successful dam project might look like.

The final report from the commission was launched in a blaze of publicity in London in late 2000 under the benign gaze of Nelson Mandela and the rather sterner visage of the World Bank's president, James Wolfensohn. It was even more scathing about large dams than the bank could have feared. It endorsed many of the environmentalists' most trenchant criticisms. Most dams didn't deliver as advertised. Average cost overruns were 56 percent. Half of hydroelectric dams produced significantly less power than promised; a quarter delivered less than half as much water to cities as their brochures claimed. Dams built to irrigate fields were no better. A quarter of them irrigated less than 35 percent of the land intended. Even dams that promised to protect against floods "have increased the vulnerability of river communities to floods," often because their reservoirs have been kept full to maximize hydroelectric production.

Dams, it said, have taken huge amounts of land—land on which people once lived and farmed. All told, at least eighty million rural people worldwide had lost their homes and livelihoods, the commission found. The Akosombo on the Volta in Ghana expelled eighty thousand people, the Aswan High in Egypt a hundred twenty thousand, the Damodar in India ninety thousand, the Kariba in southern Africa almost sixty thousand, and the Tarbela in Pakistan more than ninety thousand. Subsequently, the Three Gorges Dam displaced more than two million people.

For what? My own estimates show that the Aswan High Dam on the Nile generates just 2 kilowatts of electricity for every acre of land flooded; the Kariba Dam on the Zambezi in southern Africa generates just 1.2 kilowatts, and the Akosombo 0.36 kilowatts. Others are even worse: the Kompienga Dam in the West African state of Burkina Faso generates just 0.28 kilowatts per acre, and Afobaka in Suriname in South America is worst of all, at 0.08 kilowatts.

Dams' ecological destruction has been extensive. The commission found that far from "greening the desert" as promised, many dams have encouraged its advance by desiccating wetlands and delivering salt to fields. A quarter of the world's irrigated land, much of it watered by dams, has been damaged by salt and waterlogging. Meanwhile, dams have accumulated silt in their reservoirs—enough behind many older dams to reduce their storage capacity by more than half. And by stopping the flow of silt downstream, they have reduced the fertility of floodplains and "invariably" caused erosion of riverbanks, coastal deltas, and even distant coastlines. Coastal lagoons are being washed away all along the West African coast, owing to dams thousands of miles away. By interrupting natural river flows, dams have wrecked fisheries from the Columbia to the Ganges. Fish stocks established in the reservoirs behind dams rarely come close to compensating.

The commission report trashed some of the most treasured icons of the industry. Typical was the billion-dollar Kariba Dam, built by the British on the Zambezi in 1959. It created what was at the time the largest artificial lake in the world, on a rich floodplain where fifty-seven thousand Batongan people had lived. It was intended to generate hydroelectricity to boost the future independence and economic development of the peoples of Zambia (then called Northern Rhodesia). But far from benefiting from the project, the Batongans found themselves expelled and left destitute in refugee camps, while the electricity and water from the dam went to multinational corporations running copper mines. Was this development? The commission doubted it.

It also cataloged how the Manantali Dam on the Senegal River in West Africa eliminated floods that had provided free irrigation for half a million farmers, and how the Akosombo Dam in Ghana flooded an area of fertile farmland the size of Lebanon while providing a paltry amount of power sold at knockdown price to an American aluminum-smelting company. The loss of silt downstream of the dam has caused massive erosion at the mouth of the Volta River and along the West African coastline. Some ten thousand people who lived along the coast of neighboring Togo had seen their homes disappear beneath the waves.

The costs incurred continue to grow. Dams bring diseases including malaria, because the open water sources they provide are ideal breeding grounds for mosquitoes. More than half of sub-Saharan Africa's thirteen hundred large dams are in malarial areas, with fifteen million people living close to their shores. They may cause as many as a million cases of the disease every year.[9]

Yet dams are ubiquitous; there are few wild undammed rivers left. Most are in the empty Arctic tundra and northern boreal forests. The largest surviving wild river system is the Yukon in remote northern Canada, the world's twenty-second largest river by volume. Of Europe's last four undammed river systems, three are in northern Russia and the other, in Albania, is threatened by a cascade of proposed dams. Wild rainforest rivers that have succumbed to large dams in recent years include the Salween, which runs for 1,500 miles from China through the jungles of Myanmar and Thailand; the Rajang, in Malaysian Borneo; and the Jequitinhonha, in Brazil.

15

THE NEW DAM ERA

SINCE PUBLISHING THE scathing report of its World Commission on Dams, the World Bank has slowly moved back into funding these projects. It claims that better environmental and social standards have reduced the downsides of schemes it promotes. The second decade of the new millennium also saw a re-surgence of investment in hydroelectric dams from other agencies.[1] By one esti-mate, in 2014, at least thirty-seven hundred hydro-dams were either planned or being built—enough to increase global hydro-capacity by 72 percent.[2] Around 2 percent was added to capacity in 2016 alone, according to the International Hydropower Association, which represents dam builders.

Brazil has been busy. In 2015, it finally closed the Monte Bello Dam on the Xingu River, after a thirty-year court battle with the Juruna tribe, who lost their lands as the reservoir filled.[3] The 3-mile wide hydroelectric dam, across a major tributary of the Amazon, cost $14 billion to build. It was for a while the biggest building site in South America, with twenty thousand workers. Once all eigh-teen turbines are in place in 2019, it will be able to generate 11,000 megawatts, making it the world's third-largest hydro-dam.

Far outpacing Brazil in this new era of dam building, however, is China. Chinese companies are now the world's largest dam builders, with up to a half of all contracts.[4] They are especially busy in Africa, where their portfolio in-cludes the Bui Dam in Ghana, which will flood a quarter of the Bui National Park; the Merowe Dam on the Nile in Sudan, which has displaced fifteen thou-sand families; the Fomi Dam in Guinea, which threatens havoc on the river Niger; and "Nigeria's Three Gorges," the 3,000-megawatt Mambilla Dam on the river Dongo.[5] Ethiopia is also on a dam-building spree. Along with the Gibe III Dam on the river Omo, its 6,000-megawatt Grand Renaissance Dam, on the Blue Nile near the border with Sudan, will supplant the Aswan High as Africa's biggest.

One among many major African rivers facing a hydrological makeover is the Zambezi. The river and its tributaries pass through nine countries in southern Africa. Only a fifth of its average annual flow of more than 80 million acre-feet is currently harnessed. The Itezhi-Tezhi Dam has damaged grazing on the Kafue Flats in Zambia, where a quarter-million cattle gather in the dry season. Downstream, two big hydroelectric dams—the Kariba on the Zambia-Zimbabwe border and the Cahora Bassa in Mozambique—enfeeble the flood pulse, damaging ecosystems, fisheries, flood-recession irrigation, and wet pastures all the way to the Indian Ocean.

The river's annual flood still brings rejoicing in anticipation of an abundant harvest. On the floodplain in western Zambia, the Lozi people celebrate by bearing their king on a barge to higher ground. But "new development finance is pouring into the Zambezi basin," says Kurt Jensen of the Danish Institute for International Studies in Copenhagen.[6] He compares the Zambezi's future to that facing the Mekong. Politicians, industrialists, and financiers here all "see dams as a development driver," says Dinis Juízo, a civil engineer at the Eduardo Mondlane University in Maputo, Mozambique.[7] China is helping fund the 1,500-megawatt Mphanda Nkuwa project in Mozambique. Zambia, where the river rises and which contains almost half of the river's catchment area, has the biggest plans. First up is the $6 billion Batoka Gorge Dam, which would flood 10 square miles almost to the foot of the Victoria Falls to power copper mines and export into the emerging southern African power grid.[8]

Africa's greatest river, the Congo, may also be tamed by a series of giant hydroelectric dams with a combined generating capacity twice that of Three Gorges. In 2013, the Democratic Republic of the Congo (DRC) and South Africa agreed to harness Inga Falls, a massive series of rapids, where 34 acre-feet of water a second descends 315 feet in 9 miles just downstream of the capital Kinshasa. South African hydro engineer Henry Oliver once called Inga Falls "one of the greatest single natural sources of hydroelectric power in the world." Two small schemes built in the 1970s and 1980s, Inga I and Inga II, are largely moribund, victims of the DRC's wrecked economy and long-running civil war. But all told, the new project, budgeted at $50 billion, could generate 40,000 megawatts and deliver half of the electricity currently used across Africa. Sinohydro, the world's largest dam builder, is among those in line for contracts.[9]

A treaty signed between the DRC and South Africa pledges both countries to link the megaproject to a planned southern African supergrid. With the right infrastructure, the force of the river Congo could supply electricity to Nigeria and Egypt too. The World Bank said that the project could "catalyze large-scale benefits to improve access to infrastructure services" in Africa.[10]

Opponents insist, however, that even the Congolese will lose out. In the DRC, only 9 percent of the population has access to the electricity grid, says Rudo Sanyanga, Africa director at the California-based NGO International Rivers. "Grid-based electrification is not a realistic option." Cities might gain, but "billions of dollars in aid for the energy sector will once again bypass Africa's rural poor." She backs instead investing in local solar power and small-scale hydro schemes.[11]

That is the nub. Would the tens of billions of dollars being earmarked for large dams in Africa be better spent on other ways to deliver energy and development? Has the sheer size and grandeur of large dams once again blinded politicians and development professionals to the alternatives? Two recent international studies strongly suggest that is the case: most dams are bad deals—not just for the environment or the people whose lives are immediately damaged but also for national economies and investors. They are a black hole for scarce development cash, and typically they worsen water scarcity problems, especially downstream.

In the first study, Oxford University geographer Atif Ansar, with business and statistics colleagues, found that most of the $2 trillion or so spent on building large dams around the world in the past century has been a waste of money. As many as half had a negative economic return. And the bigger the dam, the worse the outcome. The money would almost always have been better spent elsewhere.[12]

Ansar analyzed 245 representative large dams—including 26 megadams more than 500 feet high. They had been built over the past eight decades at a combined cost of more than $300 billion. Most were hydro-dams. "We find," he concluded, "that even before accounting for negative impacts on human society and environment, the actual construction costs of large dams are too high to yield a positive return." Three-quarters of the dams came in over budget—on average 96 percent over. One in ten cost three times more. The worst—the Visegrad Dam, built by Yugoslavia and now in Bosnia—ended up a hundred times over budget—and still its reservoir leaks into the surrounding limestone rocks.

On average, dam proposers promised governments a 40 percent return on their investment. But almost half ended up as economic loss-makers. An average dam took 8.6 years to build, but bigger dams took longer. And the longer they took, the more costs ran out of control. The problems range from engineering hubris and corruption to failures of project management, economic pitfalls like inflation and currency fluctuations, fraudulent science, unexpected geology—and political vacillation. They found no evidence that things were improving. Future investment is likely to continue throwing good money after bad.

The second study looked at the winners and losers of dam construction. Dams of course capture water, and that is supposed to be a good thing, securing water supplies for an estimated half of the world's population still suffering from water stress. But the question that is rarely asked is who gets the captured water. Do dams in the real world deliver water to alleviate water shortages? It turns out that in most cases, they do the opposite. They create water scarcity, especially for people living downstream. "Almost a quarter of the global population experiences significant decreases in their water availability due to human interventions" on rivers, says Ted Veldkamp of the Vrije University, Amsterdam. Her detailed month-by-month worldwide study, coauthored with hydrologists from seven nations, found that the winners were mostly upstream but that they were outnumbered by downstream losers. Some 23 percent of the world's population was left with less water, compared to only 20 percent who have gained. Moreover, the worst downstream impacts of dams happened "in those months with the highest pressure on the available water resources"—the very months when dams are supposed to deliver their greatest benefits. In bad times, dams were making things worse.[13]

Such findings should surely change thinking around the world on dam building. We shall see. Many political leaders remain unreconstructed dam-lovers. They agree with the late Ethiopian prime minister Meles Zenawi, who said of his latest hydro-dam project on the river Omo in 2011: "We cannot afford not to have Gibe III. . . . We want our people to have a modern life."[14] But if modern life means taking a hardheaded approach to economics, then the bottom line is that most dams simply do not deliver enough power to justify the cost of building them. Moreover, most dams deprive more people of water than they benefit. They empty treasuries and faucets; they impoverish nations.

This is not just the fault of dam-flaunting dictators and despots. A final disturbing truth uncovered by the Ansar study is that, if anything, dams do worse in democratic countries. "While democracies do not take longer to build large dams than autocracies, democracies appear to be more optimistic," he concluded. Politicians on the stump love promising their constituents big dams, and voters love hearing these promises. The truth, however, is that dams are a poisoned chalice. They corrupt decision-making, and the biggest ones corrupt most of all.

16

SUN, SILT, AND STAGNANT PONDS

YOU COULD FILL every faucet in England for a year with the amount of water that evaporates annually from the surface of Egypt's Lake Nasser. This is one of the more staggering statistics I uncovered while researching this book. According to Egyptian hydrologists, between 8 and 13 million acre-feet of water disappear from the surface of the great reservoir behind the Aswan High Dam—that is, around a quarter of the average flow of the river into the reservoir, approaching 40 percent in a dry year.

This huge waste of water in a country dependent for its survival on the Nile should come as no great surprise in Cairo. A century ago, British colonial engineers forcefully advised against building a dam in the Nubian Desert for precisely this reason. It would be far better, they said, to capture water farther upstream, in the mountains of Ethiopia, where most of the river's water rises. There, a cooler, cloudier climate would reduce evaporation rates, and the steep valley would reduce the surface area of the reservoir exposed to the sun. But in the 1950s, when Egypt had gained full independence and could do what it liked, its leader Gamal Abdel Nasser decided on a dam on Egyptian territory, whatever the water losses. The Aswan High Dam created the second-largest artificial lake in the world, with water spreading for 300 miles along the flat Nubian Desert, flooding the famous ancient Abu Simbel temples and much else. They called it Lake Nasser. It is the late leader's great legacy, and great folly.

Lake Nasser is not alone in such losses, of course. Right across the tropics and beyond, evaporation is a major drain on reservoirs. A typical reservoir in India loses 5 feet. In the American West, more than 6 feet of water evaporates annually from reservoirs like Elephant Butte on the Rio Grande, and Lakes Mead and Powell on the Colorado. A tenth of the flow of the Colorado River evaporates from Lake Powell alone. In the parched Australian outback, the losses can exceed 10 feet a year.

What proportion of the river's water supplies such losses represent depends on the ratio between surface area exposed to the sun and the reservoir's capacity, coupled with the amount of time the water spends in the reservoir. The Kariba Dam on the Zambezi loses 10.4 million acre-feet a year. That is around a quarter of its annual inflow, and more water than is consumed annually in Zimbabwe. Worse still is the Akosombo Dam in Ghana, which holds back a larger surface area of water than any other dam—approaching 4,000 square miles. A typical evaporation rate of 7 feet in a year means that when the reservoir is full, it loses 15 million acre-feet a year, which is more than half the average annual input from the Volta River.[1]

Igor Shiklomanov, a Russian hydrologist, estimates that global evaporation losses from reservoirs are equivalent to a twentieth of the water eventually taken from them for human use. That's 285 million acre-feet a year, of which 40 percent is lost in Asia, a quarter in Africa, and a sixth in North America.[2] In Australia, twenty million people lose something like 3.2 million acre-feet of water a year—or 53,000 gallons a head. Peter Gleick of the Pacific Institute, the author of a biennial report on the world's water, estimates that an average hydroelectric dam in the United States loses a third of an acre-foot of water per year for every person supplied with electricity.

Could this evaporation be prevented? Technologists have suggested putting an ultrathin layer of organic molecules across the surface of reservoirs that might reduce evaporation by up to a third. But on larger reservoirs, wind would probably break the protective layer. Meanwhile, the ecological effects of cutting off the exchange of gases such as oxygen and carbon dioxide between air and water remain largely unknown. The best answer may be more clever: to cover the surface of reservoirs in hot, sunny areas where evaporation rates are highest with solar panels. They would both harvest solar energy without using up scarce farmland and prevent most of the evaporation of water. After a few small trials round the world, India, in 2016, announced plans for a 600-megawatt floating photovoltaic system on Koyna Reservoir in Maharashtra.[3]

Fetid, choked with weeds, and swarming with mosquitoes, the Balbina Reservoir in the Amazon rainforest 100 miles north of the jungle city of Manaus is a billion-dollar boondoggle. The dam rises 150 feet above the forest floor on the Uatuma River, a tributary of the Amazon. Its reservoir floods an area of forest forty times the size of Manhattan, but much of it is less than 15 feet deep. Philip

Fearnside, of Brazil's National Institute for Research in Manaus, has counted fifteen hundred islands and "so many bays and inlets it looks rather like a cross-section of a human lung."

Weeds spread across the surface of this fetid mass of water. Even the introduction of a herd of grazing manatees has failed to limit them. Stagnant water takes years to slip through the reservoir's flooded forest to the dam's hydroelectric turbines. Those turbines have a paltry generating capacity of 250 megawatts. The reservoir floods the equivalent of a soccer field to deliver enough power to run an air-conditioning unit back in Manaus.

Even in the Amazon rainforest, that sounds like a waste of land, and water aplenty evaporates from the shallow reservoir's surface. But the true insanity is even greater. The vegetation from the flooded forest, rotting in the stagnant waters, has been producing huge amounts of methane, one of the greenhouse gases thought to be responsible for global warming. It bubbles up from the reservoir bed, through the fetid waters and into the atmosphere. The reservoir was created in the 1980s to provide pollution-free electricity for the capital of the Amazon. By Fearnside's calculations, the methane it produces has four times the greenhouse effect of a coal-fired power station with a similar generating capacity. This is not green electricity.[4]

Balbina is not alone. Brazil is largely powered with hydroelectricity, and Marco Aurelio, of Cidade University in Rio de Janeiro, calculated that up to half of its hydroelectric reservoirs warm the planet more than an equivalent fossil-fueled power plant would.[5] The World Commission on Dams warned that methane or carbon dioxide bubbles up from every reservoir where measurements have been made. While only a minority are emitting more than fossil-fueled power stations, almost all make a significant contribution to atmospheric concentrations. "There is no justification for claiming that hydroelectricity does not contribute significantly to global warming," the commission said.[6]

It has proved hard to figure out how big the contribution of these emissions is to global warming. Early estimates ranged as high as 7 percent of all human-caused greenhouse emissions.[7] The current best guess, based on a detailed study of over 250 reservoirs, is rather lower, at between 1 and 2 percent. That is still, as John Harrison of Washington State University points out, equivalent to the entire emissions of Canada.[8]

Is this just a brief methane belch in the weeks and months after the reservoir fills? Initially, many scientists thought so. They figured that the gases came mostly from vegetation trapped underwater when the reservoir first filled, and the rotting vegetation would soon be gone, ending the emissions. Not so, it turns out. As reservoirs age, most have continued to produce substantial quantities of methane. One reason is that rotting can be very slow. It takes up to five hundred years for a tree to rot in a stagnant Amazon reservoir. Another is that a lot of the rotting vegetation does not come from the land flooded by the reservoir; it floats down the rivers that drain into the reservoir. As long as the reservoir continues to flood, the vegetation will continue to arrive and the reservoir will continue to give off greenhouse gases.

Of course, most of this vegetation would have rotted anyway. However, without reservoirs, this decomposition would most likely occur in a well-oxygenated river, producing carbon dioxide—whereas tropical reservoirs usually contain little oxygen and, as a result, instead generate methane. Molecule for molecule, methane is twenty times more potent as a greenhouse gas than carbon dioxide. Reservoirs thus change the way significant amounts of the earth's vegetation rots, and with that, dramatically raise the greenhouse effect of the rotting.

In 2017, these reservoir emissions had not been included in national greenhouse gas inventories and are therefore not covered by the 2015 Paris climate agreement. That may change when a task force set up by the UN climate panel completes a review in 2019. If so, then the tiny South American nation of French Guiana will find its officially declared greenhouse gas emissions are transformed. Currently, French Guiana is regarded as having one of the world's lowest emissions of greenhouse gases, with a small population and scant industrial emissions. But a dam built in the jungle to power the launch site for Europe's Arianespace rocket is a methane emitter on the scale of Balbina. The Petit-Saut Dam produces three times as much greenhouse gas as an equivalent coal-burning power station. Include those in the national inventory and French Guiana's true emissions, per head of population, are three times those of France and greater even than those of the United States.[9]

Reservoirs are not permanent structures—or not permanently useful, anyway. However well they are managed, they accumulate silt. It is eroded by running water in the mountains, brought by rivers to the reservoir, where in the slow-moving waters it settles out onto the reservoir floor. The extreme case is the Yellow River, the world's siltiest river. After the Sanmenxia Dam was completed in 1960, the river's silt filled its reservoir in just two years (see chapter 17). China's dam engineers are more sophisticated about how they manage silt flows these days, but the reservoirs on the river still probably have a half-life of just a few decades.

Most other rivers flowing out of the Himalayas carry substantial silt flows that are likely to make them as good as useless in forty or fifty years.[10] Even the Three Gorges Reservoir may suffer this fate. In China, which has more reservoirs than any other country, the amount of reservoir capacity lost each year has been variously estimated at between 3 and 7 percent. As a result, according to the International Commission on Large Dams (ICOLD), China's current stock of reservoirs has so far lost about a third of its capacity. That figure is set to approach two-thirds by midcentury.[11]

Overall, ICOLD reckons the world's reservoirs are losing storage capacity to silt at a rate of up to 1 percent a year. That is a loss of several tens of millions of acre-feet a year. By some estimates as much as a fifth of the silt traveling the world's rivers ends up trapped behind dams.[12]

What should be done? In theory, clogged reservoirs could be replaced by new ones. The trouble is that most of the world's best potential sites for dams are already taken. So replacements will almost always have less water storage and less hydroelectric potential than their predecessors. They will also flood more land for less benefit—and be substantially more expensive. The main alternatives are more effective management. Dam managers in China now routinely open their dam gates to let through the siltiest water, which is usually at the start of the monsoon season. On the Yellow River, as we shall see in the next chapter, they have devised means of flushing reservoirs of their silt by making sudden releases from a dam established just for this purpose.

VI

≈≈

When the rivers run dry . . .

floods may not be far behind

17

CHINA

The Hanging River

IN EARLY JUNE 1938, at the height of the war between China and Japan, Chinese generals perpetrated what remains the single most devastating act of war ever. They did it to their own side. Fighting to hold back Japanese invaders, the generals sent eight hundred troops to the great dike that holds in check the Yellow River as it enters the wide North China Plain before flowing into the ocean. They told the troops to dig pits beneath the four-hundred-year-old dike and fill them with explosives. On the morning of June 9, the generals ordered the detonation of the explosives and the firing of artillery shells to widen the breaches. The aim was to destroy the dike and create a great wall of water to stop the Japanese army's advance.

At first, the water flowed only slowly through the half-mile-wide gap in the Huayuankou dike. It seemed as if the plan had failed. But in the days that followed, as the summer monsoon floods came downstream, more and more of the river escaped the confines of its channel and spread across the plain. By the end of June, villages, towns, and farmland across three provinces had flooded and millions of Chinese had fled their homes.

More than half a century later, I discovered hanging in a small museum near the dike a dog-eared photograph taken in July 1938 from what remained of the dike. It showed a scene of utter devastation. I have only ever seen one picture like it, taken shortly after the atomic bomb was dropped on Hiroshima in 1945. The flooding delayed the advancing Japanese for about a month. It seems small recompense for the thousands upon thousands of Chinese drowned as the Yellow River marauded across the wide plain. And as the flood swelled, famine and disease spread. Chinese historians put the final death toll at a staggering 890,000 people, virtually all of them Chinese.

Stop a while and take that number in. It is three times the number killed by the Indian Ocean tsunami of December 2004. It is greater than the combined death tolls of the atomic bombs dropped on Hiroshima and Nagasaki and the firebombing of Dresden and Berlin and Tokyo during the Second World War.

During the flood, the river abandoned its old course and chose a new path, finally reaching the ocean nearly 450 miles farther south than before, near the mouth of the Yangtze River. (Imagine, if you can, the Mississippi heading east to enter the Atlantic in Georgia, or the Rhine rerouting to flow into the Bay of Biscay instead of the North Sea.) There it stayed for almost a decade, until 1947, when engineers managed to divert it back within its old dikes. All that remains on the rebuilt Huayuankou dike today to remind visitors of the destruction once wrought there is a small stone memorial. On the day I visited, most of the people milling around it were uncomprehending young Chinese enjoying the last autumn sunshine. Occasionally older citizens stopped for a while, consumed in thought. As well they might.

The Chinese have always called the Yellow River their "joy and sorrow." Joy because control of its waters to irrigate crops has sustained more people for longer than anywhere in history; sorrow because of what happens when the river fails or the control gives way. Unlike the Mekong, where floods are still good news, flooding along the Yellow River is usually disastrous. The river has probably killed more people than any other natural feature on the earth's surface. Controlling those floods has always been the single most important activity of Chinese governments. Many historians argue that it is the single most important reason for the creation and survival over the millennia of the vast Chinese state with its draconian powers. The Chinese sum up the relationship in a word: *zhi,* which means both "to regulate water" and "to rule."

Fast-forward from the great disaster of 1938 and the news in more recent times has been all about the river running dry. The Yellow River first failed to reach the sea in 1972. Between then and 1998, it stopped short of its delta for part of almost every year. In 1997 it failed to reach the sea for more than seven months. For most of that time it trickled into the sand on its bed near Kaifeng, 485 miles inland and just downstream of the site of the 1938 breach. Since then, by judicious releases of water from reservoirs, the river's engineers have met the demand of their political masters that they keep the river flowing into the sea at all times. But only just. Flow is often perfunctory.

I set out on a journey up and down the river to discover why the river seems to have changed its character so much—why the killer floods are giving way to droughts. On the way, I began to conclude that the river has changed less than it appears. I found that the two phenomena of flood and drought on the river are inseparable, like *yin* and *yang.* And while Chinese commentators rail against the river running dry, many experts at the government's Yellow River Conservancy Commission, whose job is to keep the river flowing and prevent disasters, believe that the next great disaster could be another major flood.

"The flood threat of the Yellow River is still the hidden danger in China," the commission's director, Li Guoying, told me as I wound up my visit. Floods remained the most fearsome threat on the river. A minor breach of an embankment could cause one of the worst catastrophes in human history. The truth, he said, was that floods and droughts have always gone together here. Scarily, every year of low flows increased the chances of a future lethal flood.

The Yellow River is the fifth-longest river in the world. It begins high in the mountains of eastern Tibet. It loops first north through the Gobi Desert and then south and east, through the heart of the world's most populous nation, to the Yellow Sea, a journey of more than 3,000 miles. Almost half a billion people depend on it for water to drink and to grow their crops. Irrigation water from the river has made China the world's largest producer of wheat and second-largest producer of corn. But this success story has been threatened today by growing water shortages.

The river is running low all the way from its source to the sea. Up on the eastern edge of the Tibetan Plateau, the area where it rises used to be known as "the county of thousands of lakes." More than half the lakes have disappeared in the past thirty years, however. In the river's middle reaches, irrigation canals are running dry, fields are being abandoned, and desertification is generating huge dust storms that choke Beijing, close schools in Korea, dust cars in Japan, and rain onto mountains across the Pacific in western Canada. Near the river's mouth on the fertile expanses of the North China Plain, springs no longer gush, the meadows are gone, and few of the old lakes that once dotted the plain survive.

The desiccation of the Yellow River and its basin is an economic as well as an environmental disaster. As its waters have faltered, millions of acres of farmland in the river's lower reaches have been abandoned. What has gone wrong? For one thing, there has been meteorological drought. But an equal cause of the river's fitful flow is a dramatic rise in abstractions to feed irrigation and China's fast-growing cities. State irrigation projects dotted along the river cover 30,000 square miles and soak up the majority of the river's water.

China has put a giant bureaucracy, the Yellow River Conservancy Commission, in charge of the river. I traveled the river with its staff, constantly struck by how their comments jumped from civil service reticence to open candor. One of the problems, they said, is that "the most inefficient irrigation schemes are always in the driest places." On the fringes of the Gobi Desert—in the provinces of Gansu, Ningxia, and Inner Mongolia—it can take four times as much water to grow a field of wheat as it does in the lower river basin. Yet in many years, most of the water is taken in these dry upstream provinces and little or nothing has been left for Shandong and Henan on the North China Plain, where the soils are better, evaporation is less, and the water would be much more productive.

In an office way upstream at Lanzhou, the capital of Gansu Province, one commission official pointed to a map of local irrigation projects. One pumped water 2,000 feet up from the Yellow River onto sandy hilltops. Another sent the precious liquid through 55 miles of tunnels to reservoirs, where much of it evaporated. Was this a good use of the water? The local officials who built the projects said yes. The water grows crops—end of story. The commission's river managers said no. The irrigation projects were hopelessly uneconomical, built more for prestige than for profit, and vastly wasteful of water.

As China has industrialized, the cities have increased their water demands, threatening agriculture. Around a quarter of the flow is piped out of the Yellow River basin, much of it to distant cities. Taiyuan and Hohhot, the capitals of Shanxi and Inner Mongolia, have both built canals 60 miles long to connect to the sputtering lifeline. That is putting ever greater stress on the river.[1]

To see the impact of these follies farther downstream, I went to the People's Victory Irrigation Project. It is just over the river from the Huayuankou dike, where the river flows out onto the North China Plain. The project is a fifty-year-old Communist totem—one of Chairman Mao's first public works to feed his people. Its canals irrigate an area about the size of Greater London, where a million people live and work. But things were in a sad state. The main distribution canal was badly polluted with foam from a local paper factory. Much of the equipment had not been replaced since the 1950s. The original iron sluice gate that the "great helmsman" Mao opened more than half a century ago was still in place. Worse, the water was giving out.

In the old days, the project was literally awash with water. It sucked almost 800,000 acre-feet from the river each year. The entire project was shut down for a while because it became waterlogged. No longer, they told me. "We are always short of water now," director Wang Lizheng said with a sigh, in between nervous puffs on his People's Victory cigarette. "We are allowed to take only half as much water from the river as we used to." Shortages were made worse because more than half the water disappeared into the ground through the bottom of leaky canals. Wang's main task was finding enough cash to enable him to line the canals. At the time, he had lined only a fifth of the main canal. "We do about a mile more a year," he told me. At that rate, it would have taken till 2025 to finish the job. In any event, that was not going to change the amount of water available. Farmers were already drilling wells to catch the water that seeped from the canals, and the more canals were lined, the less seepage there would be.

Downstream from the People's Victory project, water shortages have been even greater. In a good year, the coastal province of Shandong can produce a fifth of China's corn. However, in the 1997 growing season, the river failed to reach the province. Crops died in the fields. The crisis provoked the edict from Beijing that the Yellow River should never be allowed to dry up. The government for the first time began to enforce limits on how much water each province could take from the river, and it told the Conservancy Commission to manage the river so that a minimum flow of 13,000 gallons a second always reached the sea.

In the headquarters of the commission in Zhengzhou, I watched the giant control system in action. One wall was covered with a huge electronic map of the river, with real-time readings of key hydrological data. An alarm sounded if any of the dozens of automatic monitors on the lower river detected that the flow was getting down to the limit. The operators knew who was taking water from the river and how much. They could press a button to shut sluice gates and reverse pumps to stop all abstractions. To make the point, the engineers reset the emergency shutdown limit to just above the flow on the day of my visit, so I could hear the alarm sound. The sense of power over a mighty river was palpable.

As the river empties, farms and cities alike have tried to keep going by pumping water from beneath the ground. In Shandong on the delta, more than half the irrigation water now comes from underground. The inhabitants are pumping up water twice as fast as it is being replenished by rains and the river. The underground reserves on the North China Plain are being emptied. In the 1960s, the water table was almost at the surface; now it is more than 100 feet down. On average, water tables are falling by 3 feet a year, according to a 2015 study.[2] In places around Beijing, 90 percent of the replenishable water is gone, and here and there the city is tapping water half a mile down in fossil aquifers that will never refill. As in India, so in northern China: as the rivers run dry, people are pumping out underground water.

≈≈≈

They don't call the river Yellow for nothing. Its waters are thick and silty. Standing at the Hukou Falls, in the river's middle reaches, a popular spot for Chinese tourists, I watched what looked like chocolate mousse churning over the giant boulders into a narrow gorge. Every year roughly a million tons of silt flow over these falls and on down the world's muddiest river. Every ton of water contains about 90 pounds of silt, twice the load of its nearest rival, the Colorado. The silt, like the water, brings joy and sorrow: joy because it makes the lower valley supremely fertile but sorrow because of what it can do to the river when it is not properly managed. As the river's successful rulers through the ages have always known, managing the silt is as important as managing the water.

To see this in action, I went to a region near the middle of the river, where it cuts through steep terrain known as the Loess Plateau. The plateau is the source of 90 percent of the silt in the world's siltiest river. Nowhere on earth loses as much soil to erosion. This is because the Loess Plateau is not a proper mountain range at all. There is no underlying geology. It is just a huge pile of loose sand, several hundred yards thick and covering an area five times the size of Louisiana.[3] Much of the plateau is bare, without any kind of vegetation. Every drop of rain eats away at the sand pile. During the monsoon season, the water washes in torrents down the plateau's quarter of a million gullies (that's the total so far, but only those more than a third of a mile long have yet been counted). These muddy torrents can carry almost as much silt as water.

Erosion on this scale changes the landscape beyond recognition. Some five thousand years ago, this pile of sand was a true plateau, dotted with fortresses that have since earned it the title "the cradle of Chinese civilization." But today the "plateau" has become a maze of steep-sided hills and valleys with occasional surviving flat hilltops, some still occupied by villages and fields. For China, the erosion of the Loess Plateau is an environmental disaster—both for the plateau itself and for the river downstream, where the silt clogs reservoirs and raises the riverbed itself.

The silt in the Yellow River has been a constant source of embarrassment to China's dam builders. The faces were reddest after the completion of the Sanmenxia Dam in 1960. The first dam on the main stem of the river, it was intended as another engineering triumph for Chairman Mao—simultaneously a means of regulating water for the People's Victory irrigation district downstream, of preventing floods, and of generating hydroelectricity. Mao expelled four hundred thousand people to make way for the reservoir and was so proud of it that he had its image printed on the country's banknotes. It was part of his 1950s campaign for "the mountains to bend their tops and the rivers to give way."

But this particular river did not give way. Within two years, it had filled the Sanmenxia Reservoir to the brim with mud. Chinese engineers have since learned how to avoid catching the worst of the river's silt flows in reservoirs. Even so, the twenty large reservoirs on the river hold billions of tons of the stuff.[4]

Since the Sanmenxia disaster, leaders have dreamed of eliminating erosion of the Loess Plateau to make the Yellow River run clear. They have spent half a century remaking the steep eroding hillsides into staircases of terraces, planting swaths of trees and bushes, and plugging the gullies with over one hundred thousand small dams. This work inspired another of Mao's slogans: "Let bald hills become green fields." In most of the world, ancient agricultural terraces are being abandoned. But here in the Loess Plateau they are still being built, probably in record numbers. More than a third of the fields have been terraced. It is claimed that they have kept about 3 million tons of soil out of the river annually, though the basis for that claim is far from clear.[5]

Even optimists at the Conservancy Commission doubt that it will be possible to cut silt flows much further. Massive erosion, they point out, is a natural feature of the Loess Plateau, and to try to halt it would be decidedly unnatural. The dream of making the river run clear, they say, is just that: a dream. The Chinese, brought up on the wisdom of managing the Yellow River, have a sensible idiomatic expression: "when the river runs clear." It means "never."

The story of China's love-hate relationship with the Yellow River and its silt goes back eight thousand years, to the dawn of farming on the river's floodplain. The first farmers set up here to take advantage of the smear of silt brought down from the eroding Loess Plateau and dropped onto the floodplain as the river meandered back and forth. But while the farmers grew good crops in the fertile silt, they were at the mercy of floods during the annual monsoon and a river that regularly changed course. So they started to build dikes to keep the river on a single path.

But from that point on, the river stopped spreading its silt. It left it instead on the bottom of its new permanent channel. The layers built up. Every year, the riverbed was a little higher. Coddled behind its dikes, it soon rose above the surrounding floodplain. What the Chinese call the "hanging river" was created. It was a point of no return. From then on, the dikes had to be raised ever higher to keep the river in.

The task could never be shirked. The consequences of failing became greater and greater as time passed and the river rose higher. Between 600 BC and AD 1949, the river changed its course across the North China Plain twenty-six times—roughly once a century. The regular cataclysms often brought down dynasties. In ancient times, if the river shifted ground, the emperor was thought to have lost the mandate of heaven and could no longer rule. Sometimes there were global consequences. The Black Death, which killed up to half of Europe's population in the fourteenth century, may have begun when rats that carried plague fled China after the Yellow River changed course.[6]

As recently as 1855, the river broke its dikes so decisively that it took over another channel that entered the sea 600 miles farther north. Then there was 1938. Since then, the river has resumed its ascent above its floodplain, held back only by dikes that have to be constantly raised by press-ganged peasant labor. In places the river is now some 65 feet above the land outside the dikes. Never have the human and political consequences of a breach been so great.

Why the apparent rush to disaster? One problem is the rate at which the bed of the modern Yellow River is accumulating silt. Until the mid-twentieth century, at least half of the silt made it to the sea. Since then, with less water coming down the river, and the speed of flow reduced as a result, more silt has accumulated either behind dams or on the riverbed. By the turn of the century only about 10 percent reached the sea. The riverbed was rising by almost 4 inches a year, or over 3 feet every decade. Dikes could no longer be raised to keep up, so the capacity of the river channel was getting smaller. Its ability to contain summer floodwaters was diminishing alarmingly.[7]

In 1950, the river's inner dikes could contain a flow of up to 6.5 acre-feet a second. By the end of the 1990s that was down to a measly 1.6 acre-feet. Of course, the flow in the river was also less, and river engineers made sure to empty reservoirs before each monsoon season, so they were ready to catch whatever floods came down. But this only exacerbated the problem for the long term, because the slow flow added to the silt accumulation.[8]

A crisis was approaching. A breakdown of the dikes and a breakout of the river seemed inevitable. It was something I pondered as I stood on the Huayu-ankou dike. On one side the water was surging past. On the other, eighty-seven million people were living in a flood zone as large as Louisiana. I looked at the marks on the dike showing high water levels during past floods in 1958 and

1996. The 1996 flood contained only a third as much water as the 1958 flood, but it rose 3 feet higher up the dike wall. Engineers told me that in September 2003, a flood of just 2 acre-feet a second had broken an illegal dike erected by villagers at Caiji in nearby Henan Province, inundating one hundred thousand. Ten thousand soldiers fought the breach and prevented a wider disaster.

The notion of eliminating silt flows out of the Loess Plateau is now regarded as a pipe dream. So a bold new policy has been adopted: to flush more of the silt to the sea, by using sudden releases from dams to send an artificial flood-pulse down the river. A fast and turbulent flow, even for a few days, could stop the silt accumulation on the riverbed and even begin to remove it. It was a genuinely innovative idea, never tried on such a scale anywhere in the world. The Conservancy Commission began to store water behind dams with the aim of releasing it all at once—like a giant cistern on a toilet—to flush away the sediment on the riverbed downstream. The lynchpin of the system was a specially built new dam, the Xiaolangdi Dam. The flushing was a dramatic event, as I witnessed when I saw one of the first releases during my visit.

The dramatic approach to handling the river's silt seems to be working. By 2015, engineers had successfully conducted nine flushes and removed more than 400 million tons of silt. They had lowered the riverbed for the 600 miles from Xiaolangdi to the ocean by an average of 5 feet. That meant more space within the banks to contain a bigger river flood. During my visit, engineers had told me they hoped that annual flushing could stabilize the channel's capacity at its then current 1.4 acre-feet a second. But they have exceeded their own expectations. By 2015, flushing had more than doubled the channel's capacity to 3.1 acre-feet a second.[9] That was still only two-thirds what it was half a century ago. So, could an enormous human disaster happen again? The answer has to be yes, but the risk is much reduced.

From its founding, Chinese society has been organized around the management of water. Dynasties rose and fell according to whether they could control the floods that came down the Yellow River and maintain the rich, fertile silt in which its citizens planted their crops. It is the supreme example of what some historians call a "hydraulic civilization," a centralized autocratic society built on the paramount need to harness water resources for food and tame them to prevent destruction by floods.

The struggle is encapsulated by the legend of Yu the Great, a leader credited with controlling the floods on the Yellow River four thousand years ago by dredging channels and allowing the water to flow to the sea.[10] On a ridge near Zhengzhou, not far from the headquarters of his modern successors at the Yellow River Conservancy Commission, his statue towers above the floodplain. One hand is outstretched, bestowing prosperity on his people. The other hand carries the tools with which he cleared the river. His face looks defiantly toward the river. Taming it had been a grim, remorseless task. And once undertaken, taming the river is a task without end. Standing on the dikes outside Zhengzhou, I considered Yu's legacy and imagined the huge endeavor by hundreds of millions of people over thousands of years to keep the Yellow River in check. The dikes constructed in the past fifty years alone have a volume equivalent to thirteen Great Walls of China. Controlling the great rivers of China has always required an iron imposition of central power. China remains the ultimate hydraulic civilization.

18

UNLEASHING THE RIVER DRAGON

IT WAS THE WORLD'S WORST DAM DISASTER. One hot August night in 1975, the operators of the 400-foot-high Banqiao Dam in Henan Province in central China believed that they had nothing to fear when a typhoon hit the hills behind their dam and the river Ru swelled. Their manuals said that the dam could survive a once-in-a-thousand-year flood on the river. So as water levels rose, they kept an eye out but did not feel the need to make any emergency releases of water from the reservoir.[1]

Unknown to them, however, another dam upstream was having trouble. Shortly after midnight, the upstream dam burst, emptying 97,000 acre-feet of water down the river. When the wall of water reached the Banqiao Dam, its operators were caught unawares. Their dam swiftly gave way, this time unleashing over 400,000 acre-feet of water, mud, and masonry that hurtled on farther downstream. One woman, the Chinese later reported, leapt clear with the cry: "The river dragon has come." The flood formed a wave 7 miles wide and 20 feet high. It traveled at more than 30 miles an hour, engulfing villages on the river banks along its path and crashing into the town of Huaibin with the force of a tsunami.

The death toll that night is estimated at around twenty-six thousand people. As many as ten times more were killed in the ensuing chaos and famine. Such was the cloak of state secrecy in China back in the 1970s, however, that it was years before the outside world learned of this human-caused disaster.

Most dams are built with the promise that they will capture floodwaters from the rivers they barricade. They will prevent floods. But one of the little-told stories about dams is how often they actually make floods. This happens primarily because of the contradictory hydrological requirements of the different uses to which dams are put. In order to catch floods, reservoirs need to be kept empty, whereas for most day-to-day operations, like feeding irrigation systems, providing urban water, and generating hydroelectricity, they need to be kept as full as possible. Smart dam managers try to reconcile these aims. But as the late Bryan Davies, a river ecologist with the University of Cape Town, told me after dam releases exacerbated floods in Mozambique in 2000: "The tendency everywhere is to store like hell when you can." That can spell disaster.

In times of high river flow, a dam's capacity to capture water becomes a menace. As the reservoir fills to the brim, its operators can be faced with the choice of risking the catastrophic failure of their dam if it overfills, or making emergency releases of water down spillways that will inevitably cause massive floods downstream. Often they make the fateful decision too late. The result is releases that are far greater and more sudden than would happen during natural river flooding.

Proper records are rare, but it is increasingly clear that inept dam management has contributed to a growing toll of flood catastrophes around the world. Many fear "the big one"—a major disaster in which thousands die after a dam bursts or panicking operators release a huge volume of water to save their pride and joy. Another Banqiao.

I began to realize the flood risks posed by dams while visiting the devastation left by Hurricane Mitch, which hit Central America in October 1998. It was the most destructive storm in the Western Hemisphere in two hundred years, and dumped record amounts of rain onto steep hillsides already saturated by a month of storms. Landslides and flash floods turned small mountain streams into torrents that killed at least ten thousand people. Many would have died anyway. But some died because of the extra force and suddenness of flash floods unleashed as dams across those streams collapsed.

Records of events on the night Mitch struck Tegucigalpa, the capital of Honduras, are less than complete. When I visited the country six weeks later, I met many people who described how a wall of water had rushed out of the surrounding mountains and through poor, low-lying shanty areas in the south of the city. Several hundred died in their beds. Scientists from the US Geological Survey, who went to the city to investigate what happened, told me of their concerns about the role in all this played by two dams south of the city that had burst in the hours before the shanties were engulfed. The Los Laureles Dam had broken, unleashing a flood along the Guacerique River. The Concepción Dam had overflowed, causing what the operators' log called a "high-flow release" through the heart of the capital.

Similar explanations seemed to lie behind mysterious walls of water that hit other Honduran towns that night. I visited Pespire, a town of thirty thousand people in the south of the country. Forty-one people drowned and more than three hundred houses were washed away after a river that runs through the town rose by some 50 feet within a matter of minutes. Nobody could remember anything remotely like it from the normally placid river. The explanation may be that the town was just downstream from a dam that would normally have impounded floodwaters. When I visited, the dam was physically intact, but its reservoir was largely empty. Had its operators made emergency releases to protect the dam? Did the release trigger the disaster in Pespire that night? Nobody ever admitted that is what happened, but it seems the only explanation of events.

Dams are equipped with spillways to allow rapid release of water if strong river flows threaten the structure. They should have a capacity equal to the likely maximum flow on the river. But often the spillways are inadequate, because nobody knows what the peak flow might be. A World Bank survey in India in 1995 found two dams that could cope with only one-seventh of the possible peak flows in their river. "The consequences of dam failure during a major flood would have to be described with some adjective beyond disastrous," said the report's author, William Price.[2] Perhaps he had in mind India's worst dam failure, in 1979, when the Machhu II Dam in Gujarat burst when confronted by a river flow three times the capacity of its spillway. The water rushed downstream through the town of Morbi, one of the world's great centers of ceramics production. Estimates of the number of people who drowned range from two thousand to twenty-five thousand.[3]

One of the Indian dams that Price labeled a disaster waiting to happen was the Hirakud Dam, on the Mahanadi River in Orissa. When Sir Hawthorne Lewis, the British colonial governor, laid its foundation stone back in 1946, he promised that the dam would provide a "permanent solution" to floods on the Mahanadi. That might have been true if the dam's managers had kept the reservoir empty to await floods. But they regularly kept the reservoir too full, to ensure hydroelectricity generation and supplies for irrigators. So after its construction, the frequency of serious floods in the river's delta more than doubled. In July 2001, eighty people drowned and two hundred villages were inundated when engineers made emergency releases from the dam to prevent its destruction by floodwaters. Two years later, more than sixty died after a repeat. Still nothing changed. According to one analysis, "Out of the 19 major floods experienced in Orissa in recent years, 14 were downstream of the Hirakud and nine were caused by sudden release of water from the dam." Since 2000, "there have been five major floods, and all of them have been attributed to the dam storing water in violation of the rules."[4]

The failings of Indian dams are not unusual. In 1998, the *China Daily* reported that thousands of that country's dams were at risk of catastrophic failure. Quoting unnamed dam engineers, it said that "most of the old dams built in the '50s and '60s are seriously deformed [and] are all threatened by the hidden danger of dam collapse." It suggested that politicians were more eager to build "new star dams dazzling people's eyes" than to repair old ones, which "have been forgotten." In half a century, 322 Chinese dams had failed, culminating in the Banqiao collapse.[5]

In the chaos of heavy rains and already swollen rivers, it is often far from clear where nature's influence ends and humans' begins. Even so, it seems odd that nobody has made a comprehensive assessment of the extent of disasters caused by flood releases from dams. Alerted by what I had heard in Honduras, I carried out an ad hoc survey of media reports around the world about floods during the three years after Hurricane Mitch. Anecdotally at least, dams seemed to be implicated in many flood disasters, usually as a result of engineers opening floodgates to protect their dams. Such a release in Burkina Faso drowned forty-eight people over the border in northern Ghana. Sixty died, and eighty thousand refugees needed food and shelter, after releases from Nigeria's rickety Kainji Dam on the river Niger. A hundred drownings in Mexico were blamed on releases from La Esperanza Dam. Twenty died in Syria when a dam on the Orontes River burst.[6]

In 2005, more than a hundred people drowned after a dam a thousand feet wide burst in heavy rains on the Shadikor River in western Pakistan—just two years after a previous dam on the same site had been washed away. Then more than sixty pilgrims camping on the banks of the Narmada River in western India were swept to their deaths when managers of the Indira Sagar Dam upstream released water without warning. Considered individually, each case might be put down to bad luck or poor management. Cumulatively, however, they tell a different story. When they are needed most, dams designed and promoted to prevent floods often end up creating floods.

Many aging dams are disasters waiting to happen, however they are managed. One is the sixty-year-old Kariba Dam on the Zambezi. Whenever water is released through its floodgates, the flow erodes the riverbed downstream.

Over the years, these releases have gouged holes more than 300 feet deep into the bed close to the dam wall. This has created a growing risk to the integrity of the dam itself. As many as 3.5 million people downstream could be at risk if the giant dam bursts. After years of delays, engineers were contracted in 2017 to put things right, but the work won't be done until 2020 at the earliest.[7]

Dams are vulnerable to earthquakes. The weight of the water in their reservoirs can also cause quakes, by activating dormant fault lines deep in the earth. This seems likely to have been the case with the Sichuan quake in China in 2008, which killed around seventy thousand people. Chinese seismologists had predicted this calamity might happen. When the Zipingpu Dam was under construction on the river Min, they warned that there was a major fault nearby that might be ruptured. The reservoir began filling in 2007 and the following May, a 7.9-magnitude quake occurred 3 miles downstream. In the aftermath, other researchers concluded that the quake had indeed been "triggered by the mass loading [of water] and increased pore pressure caused by the Zipingpu reservoir." The quake would have occurred anyway, but maybe not for thousands of years.[8]

Dams are also vulnerable to warfare. British bombers destroyed dams in the German Ruhr valley during the Second World War as part of an effort to cripple the country's industrial infrastructure. American bombers similarly destroyed the Sui-ho hydroelectric complex in North Korea in 1952 during the Korean War, knocking out the country's power supplies. They again wreaked havoc on water diversions and irrigation projects during the Vietnam War. And, as we saw in the last chapter, China's destruction of its own Huayuankou dike on the Yellow River was an act of war.

The sword of Damocles still hangs over many dams in war-prone regions. The threat posed by dams is one reason some Chinese scientists opposed construction of the Three Gorges Dam on the Yangtze. They believed the dam was an easy target in case of war with Taiwan. The reservoir holds back 32 million acre-feet of water. Its breach could cause a flood that would dwarf even the loss from the Huayuankou disaster.[9]

19

CHANGING CLIMATE

ON DECEMBER 5, 2015, more water flowed down Britain's rivers than on any day previously recorded. A storm that meteorologists named Desmond drenched northern England, and rivers across the country discharged a third more water than the previous maximum, causing $700 million in damage.[1] Before the month was out, another two major storms—Eva and Frank—had tracked across the country, creating what the government's Centre for Ecology and Hydrology called "extraordinary" hydrological conditions, in which many large river catchments recorded their highest-ever peak flows.[2]

The epicenter of Desmond's downpour was Honister Pass in Cumbria, where more rain fell in twenty-four hours than had previously been seen anywhere in the country—13.6 inches. Peak river flows on the Tyne, Lune, and Eden were the three highest ever recorded in England and Wales, and more than thirty times those rivers' respective averages. As these and other rivers breached their banks, some sixteen thousand properties flooded. The heavy rains also caused landslides that silted up rivers and damaged bridges that then blocked the rivers, worsening the floods.

What caused this hydrological mayhem? Coincidentally, news of the record river flows emerged just a day after a study by meteorologists at Oxford University found that human-caused climate change had increased the chances of such torrents by between 50 and 75 percent.[3] In particular, said study leader Peter Uhe, the subtropical Atlantic waters over which the storms had passed were unusually warm. This allowed the storms to hold more moisture. The analysis "supports previous indications that human-induced climate change is increasing the risk of heavy winter rainfall" in Britain, said Peter Stott, the head of climate-change attribution at the Met Office Hadley Centre in Exeter. It was a sobering reminder of what damage climate change can do even in a country as renowned for its moderate weather as Britain.

Floods, droughts, and all kinds of meteorologic events are routinely blamed on climate change. So we should be careful of simplistic attribution. Though the broad global outlines of climate change are clear enough, few things are less certain in forecasting future climate than where weather systems will end up and what that will mean for rainfall. But what is becoming clear already is that climate change is making extreme weather more frequent, and that those extremes include exceptional rainfall and floods, as well as longer and more brutal droughts. The logic is obvious enough.

Higher air temperatures will increase evaporation from the world's oceans and allow the atmosphere to hold more water vapor. By later this century, there could well be 8 to 10 percent more water vapor in the atmosphere on an average day than there is today. That is 800 million acre-feet or so extra, enough to fill twenty Niles. This will almost certainly increase global rainfall, and make individual storms more intense. Though to complicate things, as climate change shifts the trajectories of rain-giving climate systems such as Atlantic cyclones, rainfall will be redistributed. Some places will get drier. Meanwhile, the higher temperatures will also mean soils dry out more quickly. So in some places, less of the rainfall may reach rivers. How this balance plays out is not easy to calculate. The rule of thumb seems to be that dry areas will become drier while wet areas will become wetter. Some places—maybe most—will experience both more droughts and more floods.

This prognosis matters for rivers and for how we use them. The one certainty is greater uncertainty. Droughts may not end. A once-in-a-century flood may be repeated the following year. The implications for infrastructure on our rivers will be grave. Big engineering projects such as dams that are expected to provide reliable sources of water and hydroelectricity for many decades to come badly need a certainty about future flows that they cannot have. Dams will run ever greater risks of being left high and dry—or of being overwhelmed by floods that their spillways and floodgates were not designed to cope with.

The best attempt to make sense of things is a study by Yukiko Hirabayashi of the University of Tokyo. She looked for changes in risk of extreme floods—the once-in-a-hundred-year variety—based on combining eleven climate-model predictions of rainfall and estimates of the likely impacts on river flows. She found that, by the end of the century, there would likely be "a large increase in flood frequency in southeast Asia, peninsular India, eastern Africa and the northern half of the Andes." But there would probably be a reduced risk of flooding in the drought-prone American Midwest and across a large swath of Eurasia from the Rhine and Scandinavia to Turkey and the Aral Sea catchment.[4]

The US government's Scripps Institution of Oceanography estimates that reservoir levels in the Colorado will fall by a third, as declining rainfall and rising evaporation combine to reduce moisture by up to 40 percent across the southern and western United States. It predicts that the Niger, which flows through nine nations across the arid Sahel region of West Africa, faces losing a third of its water. Inland seas, lakes, and wetlands will be at special risk, because most climate models expect continental interiors to suffer the biggest losses of rainfall and the highest increases in temperature and evaporation. The Caspian Sea and Lake Balkhash in Central Asia, Lakes Chad, Tanganyika, and Malawi in Africa, and Central Europe's largest lake, Lake Balaton in Hungary, are all on the endangered list. In contrast, the great Arctic rivers of northern Canada and Siberia will probably gain water as warmer air holds more moisture and more rain falls on their catchments. So the Mackenzie and the Yukon in Canada and the Ob, Yenisei, and Lena in Siberia will rage even more fiercely—40 percent more, according to one study.

There is some evidence that these trends are already well under way. The proportion of the earth's land surface suffering very dry conditions rose from 15 percent to 30 percent at the start of the twenty-first century. Much of the Middle East seems to be drying out already. In 2017, below-average rainfall had persisted for more than a decade in the region. Some researchers argue that empty irrigation canals and dried-up wells were the underlying reason behind the social discontent that unleashed the civil war in Syria in 2011.[5] In Iraq, less rainfall, combined with water diversions, has reduced the flow of both the Tigris and Euphrates by more than 40 percent in recent years.[6] Climatologists predicted in 2009 that the drought is likely to be permanent and the two rivers, which have sustained the region for thousands of years, "will disappear this century."[7]

Nothing is certain, however. Take the Nile. Egypt is heavily dependent on the river for both electricity from the Aswan High Dam and irrigation of its desert farms. It is already jittery because upstream Ethiopia is building a giant new hydro dam (see chapter 23). The two regional superpowers, along with the other nations on the world's longest river, have for years been trying to reach a deal for sharing its water. Any deal will have to make assumptions about future river flows. Yet nobody has a clue what they might be. One study in 2017 predicted the Nile could see an average increase of 10 percent this century.[8] But the Intergovernmental Panel on Climate Change has reported that different model projections for the Nile flow later this century range from 30 percent more water to 78 percent less.[9]

In an era of climate change, many people—including the governments of Brazil, Egypt, Ethiopia, and China—have proposed building more hydroelectric dams as a source of low-carbon electricity. The problem is that climate change itself could upset these plans. Future droughts run the risk of leaving nothing to drive the turbines. Hydroelectric dams have a design life of eighty years and more. So betting on river flows remaining secure over that timescale is high-risk.

Nowhere is safe. Michelle van Vliet of the International Institute for Applied Systems Analysis has crunched the numbers to see what standard climate models predict for more than twenty-four thousand hydroelectric power plants around the world. She concludes that up to three-quarters will see reductions in generating capacity by midcentury. Disturbingly, the threat is greatest for fast-growing economies that rely on hydropower for more than half of their electricity, notably in Latin America.[10]

The trend may already be under way. Recent unexpected droughts in Venezuela and Brazil have dried up reservoirs and caused power outages.[11] Victims of recent climate change litter the landscape. British colonial engineers seem to have been especially overoptimistic, or unlucky. In Ghana, the Akosombo Dam has been left high and dry by reduced flows down the Volta River. The dam was designed by British colonial administrators, at a time of high rainfall in the region, to kick-start industrialization as Ghana became independent in the 1960s. Instead, it virtually bankrupted the nation. These days the reservoir is rarely more than half full. Its reliable electricity output is a fraction of that anticipated. Industrialization never happened. Likewise the Victoria Dam in Sri Lanka, which was built by the British in the 1970s to industrialize a newly independent nation. It too has languished half-used for much of its lifetime because—whether or not thanks to climate change—the Mahaweli River has only seen 40 percent of the flow anticipated by British hydrologists.

Many rivers depend for much of their flow on melting glaciers. They face an especially complex future. As the glaciers melt, flow will increase, especially in summer. Then as the glaciers start to disappear, flows will decline, or alter in their seasons to reflect rainfall rather than snowmelt. This is already happening in Europe, where some 3,800 glaciers are set to disappear.[12] The Alps have lost a quarter of their ice, with the melting contributing to unprecedented floods in central Europe. But soon, as the glaciers disappear, summer flow may decline by up to 40 percent. In Canada, British Columbia and Alberta are home to nearly a tenth of the world's glaciers. They are currently losing almost 1 percent of their volume every year, says Garry Clarke of the University of British Columbia in Vancouver. This rapid melting will swell rivers between 2020 and 2040, creating bounty for hydroelectric dams such as the 790-foot-high Mica Dam on the Columbia River. We should not be misled by the big flows, however. "They will be a farewell message from the glaciers, not a sign of abundance."[13]

The glaciers of the Himalayas and the Tibetan Plateau feed seven of the greatest rivers in Asia—the Ganges, Indus, Brahmaputra, Salween, Irrawaddy, Mekong, and Yangtze. As they melt, many rivers will swell dangerously. "But once the glaciers go, you're down to whatever happens to fall out of the sky," says Martin Price, a British glaciologist at Perth University in Australia. The Indus, for instance, is expected to lose 38 percent of its flow—and 58 percent in drought years, according to Hamish Pritchard of the British Antarctic Survey.[14]

Such numbers remain extremely uncertain. Many of the predictions about rainfall changes, drought durations, flood intensity, and glacial melt assume the world does nothing to fight climate change, which after the Paris Agreement is unlikely. Equally, they assume that the climate models are right in every detail. So, while the precise future for the world's rivers in an era of climate change is likely to remain unclear until it happens, the presumptions of the past are assuredly now wrong. We just don't know *how* wrong.

VII

When the rivers run dry . . .
we go to war over water

20

HONDURAS

Berta Vive

THEY CAME FOR HER LATE ONE EVENING. As Berta Cáceres prepared for bed, a heavy boot broke the back door of the safe house she had just moved into. Her colleague and family friend, Gustavo Castro, heard her shout: "Who's there?" Then came shots. Castro survived, but the most fearless social and environmental activist in Central America, famous for battling dam projects in the mountains of her homeland, died instantly. She was forty-four years old. It was a cold-blooded political assassination.

Berta Cáceres had been fighting a flurry of dam projects launched since a military coup in Honduras in 2009. Many of them were on the traditional lands of her people, the Lenca, who in pre-Columbian times shared much of the region with the more-famous Mayans. She had known she was likely to be killed for her opposition to these projects. Everybody knew. She had told her daughter Laura to prepare for life without her. The original citation on her prestigious Goldman Environmental Prize, awarded in the United States in 2015 for her anti-dam campaigning, had acknowledged the continued death threats, before adding, in a note later removed: "Her murder would not surprise her colleagues, who keep a eulogy—but hope to never have to use it."[1]

"I knew she was afraid," said Maria Santos Dominguez, who lives in the remote indigenous village of Río Blanco, nearby one dam project. "It was too much for her. I could tell." In early 2016, it happened.

The human rights group Global Witness says Honduras is "the deadliest country in the world to defend the natural world." Cáceres had opposed a rash of development projects and concessions handed out to companies for dams, mines, and other projects. But most believe the motive for her murder was her campaign against the Agua Zarca dam project on the Gualcarque River, which flows through the beautiful mountains of western Honduras.[2]

When I visited in early 2017, prosecutors in the Honduran capital Tegucigalpa told me that eight people had been arrested in connection with the death of Cáceres. Six of them had links to government security services, including an elite military squad trained by US Special Forces. Two had alleged links to the Honduran company behind the Agua Zarca project, Desarrollos Energéticos SA. They included its former security chief and the individual in charge of environmental policies at the dam. The company was not commenting.

Cáceres's political tenacity came from generations of family politics, particularly among the women. I visited her eighty-five-year-old mother, Austraberta Flores, with whom she lived until her final months. Thrice mayor of La Esperanza, a picturesque city in the Lenca heartland, Flores had also been a congresswoman in Tegucigalpa. She promoted into Honduran law an international code requiring that communities like the Lenca are able to give, or withhold, their "free, prior, and informed consent" before development projects such as dams can go ahead on their lands. "It's still the strongest law we have," she told me proudly. "Berta grew up with struggle. She saw it every day. It was her schooling. I knew she would be important. I was always pushing her to become what she became."

The next day I traveled to Río Blanco, the distant village that became the focus of Cáceres's campaign against the Agua Zarca dam project. It had been a violent and bitter struggle. In 2013, local activist Tomás Garcia had been shot dead by soldiers during protests at a construction camp set up by Chinese engineers working on the dam.[3]

María Santos Dominguez, local leader of opposition to the dam, had also been viciously attacked and had a nasty scar on her face. She told me how the village had become divided between those for and against the dam. One pro-dam family had complained that Dominguez "spoke too much." It was her fault, they said, that they couldn't get economic development in the village. "One day, they saw me go past on the way to my children's school, and they hid for my return. Then they attacked me with a machete. I had taken my phone out to talk to my husband. He heard it all and came running. He had my son with him and told him: 'They are killing your mom.'"

Dominguez spent six months recuperating. But now she was back home, as determined as ever. She broadcast each week on the Lenca radio station, from a location that was kept secret to prevent attacks. She took me to the river. We walked down a road constructed by the dam builders. We came to a gorge where a quiet pool had formed between two rapids. It was a beautiful spot and, in engineering terms, I could see it was an ideal place for the dam the villagers and their allies had so far been able to prevent.

Dominguez bathed her children in the clear, cool mountain waters. "The river is sacred to us. We believe in the spirits in the river—they are three little girls, and they give us strength to fight the dam builders," she said. It had become an existential fight. "We were born here. It is our land and our river," she said. "If we lost the river, we'd die. We need its water to bathe, for fish, for water, for our crops and animals."

A few days before her murder, Cáceres had come to Río Blanco. "There were dam people on the river, working with machinery. It looked like they might be about to begin work on construction. So we went to see them," said Dominguez. "But they accused her of stirring us up, and they threatened to kill her. A few days later she was dead. The dam people haven't come back to the river since."

Will the Agua Zarca dam ever be built? Some now doubt it. After the outcry over Cáceres's death, the development banks of the Dutch and Finnish governments pulled out of the project. The Chinese engineers are gone too. Elsewhere on Lenca territory, however, dams are going ahead. In La Paz province, the Lenca have been fighting a rash of hydroelectric schemes on mountain streams. Most have been promoted by a local right-wing politician, vice president of the Honduran Congress, Gladis Aurora López. The Lenca say the mountains have been illegally taken from them. "They call us stupid Indians," La Paz activist Margarita Pineda Rodriguez told me. "But these projects offer us no benefits, only loss of our natural resources."

"We are seeing the recolonization of our country," said Tomás Gómez Membreño, who has replaced Cáceres as head of the Lenca campaigning organization COPINH. "More and more of our natural resources are being handed out to foreign corporations. There is more and more repression of people who fight back."

Their tenacity in the face of continuing violence remained remarkable, however. Less than two weeks after the 2016 assassination of Cáceres, another COPINH activist, Nelson García, was shot dead outside his home south of La Esperanza. Days after I visited Honduras, paramilitaries in neighboring Guatemala shot dead seventy-two-year-old Sebastian Alonso, an activist opposing a 25-megawatt hydroelectric dam at Pojom in the Sierra Madre in the west of that country, the latest of three slayings of protesters.[4]

Cáceres's modest bungalow stood empty. The only sign of her violent death was a dent in the wire fence where the assassins had climbed over. A BERTA VIVE poster hung in the window. Some activists wanted her home to become a museum of her life, to seal her martyrdom. The river she died to protect may or may not get dammed. But the battle for her legacy—and for the future of the Lenca and their rivers—goes on.[5]

21

PALESTINE

Poisoning the Wells of the West Bank

AHMAD QOT WAS A POOR Palestinian farmer. He spent between three and four hours every day walking his donkey through the hills of the West Bank to get water for his nine children and five farm animals. I met him by chance as he stood over a shaft to an ancient tunnel that channels water from the hills above Madama, his village outside the town of Nablus. He repeatedly dropped his bucket down the shaft and hauled it up, gradually filling a long line of plastic containers with water. His donkey waited patiently to be laden with the containers. "I come here three times a day," Ahmad said. "I have three cows and two sheep; this is the only water for them."

His life was breathtakingly spartan. He had less than 2 acres of land, he told me. He grew a little wheat, to make bread, and some vegetables. His sheep provided his family with occasional meat. His cattle produced milk, some of which he sold in the village. The lifeline for his animals was a shaft in the road—a shaft with a beautifully made stone surround that had been deeply grooved by centuries of ropes hauling buckets. Centuries of people getting by, as Qot was doing.

The villagers said that the shaft and tunnel were Roman, which meant little more than that they are very old. The tunnel captured seeps from tiny springs in the hills and brought their flow to the village. There are similar tunnels all across the West Bank. Hydrologists call them spring tunnels. They are an ancient water-catching system that even many Palestinians know little about. A lot have dried up as water tables have fallen, but the tunnel at Madama ran all year. Qot and a handful of the poorest families in Madama still used it to water their animals. Though lately, he told me, it had become polluted. "Sometimes even the donkeys won't drink the water," he said. Certainly his family could not drink it. So most days Qot made another, much longer journey out of the

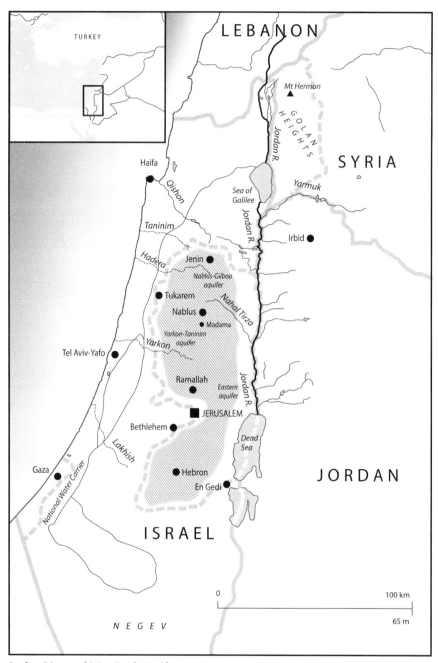

Jordan River and West Bank Aquifer

village, across a road that led to a nearby Jewish settlement and past a military checkpoint to a clean spring in the neighboring village of Iraq Burin. The journey took two or three hours—much longer if the soldiers held him up. "Some days the soldiers let me pass with my donkey, and some days they don't," he said. "Sometimes when I am coming back, they just pour the water on the ground. They never say why. Once they just drove a tank over my water containers." He could not understand such behavior. "Water is life," he explained.

Qot had no choice but to make the journey, though. He could not afford the two dollars for a thousand liters (265 gallons) charged by the driver of the village's water tanker, who also collected water from the spring in Iraq Burin and delivered it to the wealthier inhabitants of Madama.[1]

The reality of daily life for many Palestinian villagers on the West Bank is told in such stories. It doesn't explain the continuing poisonous dispute between Jews and Palestinians in these hills, but it does go to the heart of the daily misery the dispute brings.

The two thousand inhabitants of Madama used to have another source of water. There was a spring two-thirds of the way up another hillside toward Yitzhar, an Israeli settlement that overlooked the Palestinian community. A pipe connected to the spring took water down to a tank outside the Madama mosque. This had been the village's main source of clean and free water until the spring became polluted. Everybody in the village assumed that the pollution came from the Israeli settlement—and I could see no other likely cause. Neither could Oxfam's Geoff Graves, my guide in the village. "We have twice repaired the head of the spring after Yitzhar settlers damaged it. The last time we fished soiled diapers out," he said.[2]

The people in Yitzhar had a reputation as one of the more militant of the Israel settler communities, clashing with Israeli security forces as well as their Palestinian neighbors.[3] Ayed Kamal, the head of the Madama village council at the time of my visit, told me: "Sometimes the settlers shoot at us when we go up the hill. Three of us have had bullet wounds. They killed a donkey." Graves said that the settlers had even shot at Oxfam workers repairing the spring.[4]

In Madama, getting water was a constant problem. Most families collected the winter rain from their roofs and stored it underground in cisterns. "But this water doesn't usually last beyond about May. After that, we all have to buy water delivered by tankers if we can afford it, or bring it by donkey if we can't," Kamal said. But sometimes they couldn't even pay for water. The Friday before my visit, a curfew imposed by soldiers on the settlement road had prevented the water tanker from coming in.

Meanwhile, on the ridge in Yitzhar, there was running water at all times, sprinklers in the settlers' gardens and irrigation in their fields without limit. It felt like hydrological apartheid.[5]

The West Bank has few permanent rivers, but the dolomite hills are honeycombed with caves and crevices, where rainwater collects in three aquifers. The largest is beneath the western slopes and drains toward Israel proper and the Mediterranean. The second, around Nablus, drains north and supplies much of Galilee. The third drains east to the Jordan Valley. Collectively, they are called the mountain aquifers. Tapped by wells or seeping from springs, they are virtually the only source of water for Palestinians, and they are at the heart of the conflict between Israelis and Palestinians over water.

In the 1950s, when the Palestinians lived here under Jordanian rule, the West Bank seemed to have ample water. Far more rain was falling and filling the aquifers than the Palestinians needed. The excess water in the western aquifer gushed from springs along the boundary between the West Bank and Israel. Those springs were the sources of two of Israel's largest rivers, the Yarkon and the Taninim, which flowed west into the Mediterranean. As the nation's population grew, however, the Israelis began taking water directly from the western aquifer by sinking wells close to the border. Soon they were taking far more of the water from under the West Bank than the Palestinians ever had, without actually setting foot on the West Bank.

One man who helped tap into the western aquifer for Israel was Ze'ev Golani, a water engineer. "By the early 1960s, we were taking about 240,000 acre-feet from the aquifer and the Arabs about 16,000. Between us we were fully utilizing the aquifer, plus a little more," he told me. Then the water table in the western aquifer began to fall. "So, when Israel took control of the West Bank after the Six-Day War in 1967, we said there should be no additional pumping by anyone—certainly not for agriculture." That remains the situation half a century later. Israelis took the lion's share of the western aquifer and, in the name of conservation, decreed that they should take the lion's share forever.

Israeli hydrological rule on the West Bank since 1967 has been absolute and unyielding. Israel refuses to allow new wells to be dug by Palestinians, even when the old ones give out. Ironically this iron hand has been sustained by the terms of the Oslo Accords, signed in 1993 by Israeli and Palestinian leaders. The accords were intended as a temporary arrangement until a final agreement on land and water was reached. They largely reflected the status quo, giving Israel control of 80 percent of the aquifer, including the whole of the western side. But with no sign of a final agreement, the Palestinians are stuck with a blatantly unfair share-out of water.

Haim Gvirtzman of the Begin-Sadat Center for Strategic Studies of Bar-Ilan University says this mischaracterizes the situation. He claims that, despite the bans, Palestinians have sunk around 250 "illegal" wells into the western aquifer—forcing some Israeli farmers to reduce their agricultural plantings. He points out that the Oslo Accords do give Palestinians the right to drill for water at forty sites in the eastern aquifer. But he says that they have drilled only a third of them. "If the Palestinians were to develop and drill all these wells, they could have completely solved the existing shortage," adding 40,000 acre-feet to their water supplies, he wrote in 2014.[6]

Independent hydrologists I have spoken to disagree. The eastern aquifer has only about a quarter as much water as was assumed when the accords were signed, said Clemens Messerschmid, a German hydrologist from the University of Freiburg who has worked with Palestinians on their water supply for two decades. "And the only good places for drilling . . . are down in the Jordan Valley. It's just not economical. The water table is a third to a half a mile down, and there have been so many dry boreholes that at a thousand dollars a yard, the aid agencies don't want to drill anymore." The US government's aid agency, USAID, agreed. Alvin Newman, its head of water resources, told me before his death: "We are getting to the end on the eastern aquifer. If you are pumping water half a mile out of the ground and then another two-thirds of a mile uphill to faucets in Nablus, that involves very big costs."

As the existing West Bank springs and wells deteriorate and their population grows, Palestinians find they have less water per capita than when the Israelis invaded, and, at around 18 gallons per person per day, less than a quarter as much as their Israeli neighbors. The World Health Organization has warned that poor water supplies are to blame for rising rates of waterborne diseases among Palestinians on the West Bank, as they are in Gaza.

The problems have been made worse by Israel's security fence. Israel said it planned the fence to keep suicide bombers out of its towns during the intifada between 2000 and 2005. But hydrologically, it could not be more divisive. It has cut off dozens of Palestinian villages from their wells.[7] Messerschmid calculates that wells and springs on the wrong side of the barrier represent almost a quarter of the water the Palestinians once took from the western aquifer.

Of course, the Israelis have a right to security—and water. The Palestinians have rights too, however. And far from relinquishing control of the western aquifer to the people who live above it, Israel is tightening the West Bank faucet. Unless that changes, then Ahmad Qot will go on dropping his bucket into the Madama shaft for many years yet—and the prospects for hydrological peace will remain dim.

22

RIVER JORDAN

The First Modern Water War

IN 1964, Israel hijacked the waters of the Jordan River. There is no other way of putting it. The Jordan Valley, a green desert strip that had been cultivated for longer than perhaps any place on earth, was overnight deprived of most of its water. One day the Jordan poured out of the Golan Heights, into the Sea of Galilee, and on down the valley to the Dead Sea, as it had for millennia. The next day, a dam blocked its outflow from the Sea of Galilee, and a pumping station lifted the water into a 10-foot-wide pipe that delivered it the length of Israel.[1]

Grabbing the Jordan's flow was no easy task. Engineering heroics were involved. The Sea of Galilee is more than 600 feet below sea level, in a great fissure in the earth's crust known as the Rift Valley. Diverting that water to faucets as far as Tel Aviv, the Negev Desert, Jerusalem, and West Bank settlements required lifting it 1,200 feet. The pipe is known today as the National Water Carrier. It can carry more than 400,000 acre-feet a year and has for half a century been the major source of water for all of Israel.

The audacious water grab also caused a geopolitical earthquake. It was done without the agreement of either Syria, where the river rises and which then held the eastern shore of the Sea of Galilee, or Jordan, through which the river until then ran to the Dead Sea. It was as if France had one day unilaterally annexed the Rhine as it flowed out of Switzerland and pumped its contents over the hills to irrigate the plains of northern France, leaving a dribble of water to flow down Germany's main artery and into the Netherlands.

Almost three years after the hijack, Israel and its Arab neighbors fought the Six-Day War. Most histories of this conflict discuss its importance and motivation in terms of land and security. They often ignore water. Yet the simple fact is this: before the war, less than a tenth of the Jordan River's basin was within Israel's borders; by the end, the basin was almost entirely controlled by Israel.

During the brief conflict, Israel had taken from Jordan all the land between its former eastern border and the river. It had also taken from Syria the Golan Heights, the mountains northeast of the Sea of Galilee where the river rises.[2]

Ariel Sharon, who was a commander in that war and much later became Israeli prime minister, was unabashed about Israel's hydrological motives in that conflict. In the early 1960s, he wrote in his autobiography *Warrior* that Syria had begun the water war by starting to dig a canal in the Golan Heights to divert the headwaters of the Jordan away from Israel. And that "the Six-Day War really started on the day Israel decided to act against the diversion of the Jordan. . . . While the border disputes were of great significance, the matter of water diversion was a stark issue of life and death."[3]

The Six-Day War was, by this account, the first modern water war. Israel's seizing of the Jordan River and its catchment remains an essential backdrop to the continuing impasse over the region's future. Israel's victory in the war and its annexation of the river has enabled it to consume far more water than falls on its territory. It has been able to do so because of its occupation of the West Bank, which gives it control of the mountain aquifers, and of the Golan Heights, which in turn gives it control of the Jordan River. Hardline politicians in Israel see this new state of affairs as permanent.[4] But others say that almost any final peace settlement with Israel's neighbors will require the country to hand back some of its control and some of the water.

Israel now has four main sources of water. The first, used from the day the state was founded after the Second World War, is an aquifer that stretches along the coast from Haifa in the north to Gaza in the south. The Israeli water engineer and administrator Ze'ev Golani drilled many wells into the coastal aquifer as the new state established plantations of water-guzzling crops such as cotton, tomatoes, avocados, and the ubiquitous Jaffa oranges. Soon, he told me in an interview in 2004, the water table in that aquifer fell and the porous rocks began to fill with salty seawater flowing in from the Mediterranean. Pumping was cut back. To compensate, Israel began to sink wells farther inland, near the Yarkon and Taninim Springs. Those wells tapped the country's second major source, the western aquifer beneath the West Bank. By the early 1960s, the Israelis were overpumping there too. That was when they decided to create a third source, by annexing the Jordan River and pumping it into the National Water Carrier.

The fate of the coastal aquifer—much of which remains too salty for drinking—is essentially a private matter for Israelis. The fate of the western aquifer remains a matter of intense dispute between Israelis and Palestinians. What happens to the Jordan involves many more players, including Syria, Jordan, the Palestinians, and, from time to time, Lebanon. So much hangs on a river that, in strictly hydrological terms, is so small. From its sources in the Golan Heights to its former end point in the Dead Sea, it is barely 200 miles long. It drains an area the size of New Jersey and, as Mark Twain once observed, "is not any wider than Broadway in New York." But throughout human history its waters have been vital in this region, and the current battle over who controls them is in deadly earnest. Here, when the river runs dry, war looms.

In all real senses, the Jordan River is now two rivers. One begins amid the snowy slopes of Mount Hermon on the Golan Heights, which is legally in Syria but since 1967 has been occupied by Israel. Its main source here is the Dan River, which bubbles to the surface in a spring amid glades of laurel, ash, and fig close to the ancient Phoenician city of Laish. Two other streams, the Banias and the Hasbani, which begins life in Lebanon, join before it crashes down through a narrow gorge into the Rift Valley and the Sea of Galilee and then passes into the netherworld of the National Water Carrier all the way to the Negev Desert town of Mitzpe Ramon, a desolate hilltop outpost trying to rebrand itself as a tourist town. Its water is at this point 3,600 feet higher than when it left the Sea of Galilee.

The second Jordan River is the former tributary known as the Yarmuk River, which rises in Syria and enters the bed of the Jordan 6 miles downstream of the Sea of Galilee. The bed is empty here, except for a trickle of saline water from salty springs on the shores of the sea, which Israel diverts to prevent it polluting its carrier water. For many years, the Yarmuk disappeared to nothing as farmers in southern Syria erected small dams that grabbed half its flow for irrigation, while Jordan diverted most of what was left into the 60-mile-long King Abdullah Canal, which passes down the Jordan Valley on the east side of the river, irrigating fields as it goes. But the Syrian civil war resulted in farmers deserting their fields and, at the time of writing in 2017, much more water flows out of Syria.[5] Even so, the spot where John the Apostle is supposed to have baptized Jesus has in recent years been little more than a sewage seep behind the barbed wire of an Israeli military fence.[6]

And the Dead Sea? It is famously the saltiest large body of water on the planet. And the lowest. With next to no water entering it from the Jordan Valley, it is slowly evaporating in the desert sun, becoming saltier as it does. In 2017, its surface was 1,412 feet below the level of the world's oceans—44 feet lower than when the first edition of this book was published. A third of the sea's surface area has gone, and it has divided in two. The retreating shoreline is leaving behind mud and quicksand, which makes it unsafe to walk along. Thousands of sinkholes, some bigger than a car, have opened up as salt layers dry out along the shore and cavities collapse. Roads are buckling and bridges are giving way. An area around the town of Engedi has been effectively abandoned.

Israel continues its search for more sources of water. It is making much greater use of its first three sources, by recycling sewage on a scale never attempted anywhere else. Today more than 80 percent of Israeli urban effluent is captured, cleaned up, and sent to irrigate farms. Meanwhile, the country has also embarked on the creation of a fourth source of water: desalinated seawater. It has opened three giant desalination plants on the shores of the Mediterranean. Ashkelon opened in 2005, Hadera in 2009, and Sorek in 2016. Together, they deliver more than 50,000 acre-feet a year, adding a third to the nation's previous water supply.[7] The energy cost of desalinating seawater is high but not much higher than pumping water up from the Sea of Galilee.

Israel's desalination works and sewage-recycling projects mean that for the first time the nation has more freshwater than it needs, even in times of drought. In 2013, it agreed for the first time to divert some of its cleaned-up sewage back into the river Jordan. It was only effluent, and only the equivalent of about 1.5 percent of the river's former flow. But it was a start.

"We hope they will deliver more in future," said Gidon Bromberg, founder of the NGO EcoPeace, formerly Friends of the Earth Middle East. He hopes one day to see 350,000 acre-feet of recycled sewage effluent flowing into the Jordan a year.[8] That would still only be a third of its former flow, but it would allow the return of the semblance of a normal river.

I talked about all this with Arie Issar, a man steeped in the history of Israeli water. More than half a century ago, before there was a Jewish state, his father sold ice with Arab friends in Jerusalem. As a young scientist in the new state of Israel, Issar helped to green the Negev Desert by finding new sources of water beneath the sands. He says there are still more to be found, though he has always preached careful use and sharing of the region's water.[9]

We met one hot summer's day in Jerusalem. We looked out over the arid Jordan Valley, which his fellow hydrological engineers had emptied, and north to where Israeli soldiers prevented Palestinian villagers from sinking wells to meet their most basic human needs. An old Zionist idealist, Issar clearly regretted the new belligerence of modern Israel. He confided that winning the Six-Day War had been in some respects a disaster for his people: "After that, we stopped developing the Negev, which was largely empty, and began to occupy the West Bank, which was already populated by the Palestinians. Now we are fighting over land and water in the West Bank when we could be developing the land and water in the Negev. We need to get back to that idea.

"People talk about water wars, but water can also be the basis for peace," he said. "And I think it can be so here. We Israelis use too much drinkable water for irrigation when farming is no longer important for our economy. We do crazy things like turning freshwater into oranges and exporting them. The Arabs need that water. They should have it."

23

EGYPT

Gift of the Nile

FOR THOUSANDS OF YEARS, Egyptians have depended on the waters of the Nile. Flowing out of the Ethiopian highlands and central Africa, the Nile is the world's longest river, at 4,200 miles. Its drainage basin covers 1.2 million square miles and occupies a tenth of the African continent. The river passes today through eleven countries. Without its waters, the ancient civilization of Egypt, the most downstream of those nations, would never have arisen, and its current population of approaching ninety million people would have few means of survival in a barren desert. No wonder Egypt is famously described as "the gift of the Nile."[1]

So things looked bad in 2011 when, 1,200 miles upstream, Ethiopia began to build a giant hydroelectric dam across the river's largest tributary, the Blue Nile. The project was started without warning, while Egypt was in the midst of the political convulsions of the Arab Spring. When, in 2013, engineers began diverting the river's flow to begin construction of the dam itself, tensions between the two countries rose.[2]

Once completed, the Grand Ethiopian Renaissance Dam (GERD) will be the largest hydroelectric dam in Africa—and an existential threat to Egypt, able to cut off the country's water lifeline at a single stroke. Construction takes place in a legal limbo. The only existing treaty for sharing out the waters of the Nile was drawn up by the colonial British in 1959. The Nile Waters Agreement allocated 45 million acre-feet of the river's annual flow to Egypt and 15 million acre-feet to Sudan. It gave no water rights to Ethiopia at all. Egypt has always said it was determined to hold onto those rights. Boutros Boutros-Ghali, Egypt's former foreign minister and subsequently secretary-general of the UN, famously warned that his country would go to war to stop a dam on the Nile in Ethiopia.[3]

Thankfully that hasn't happened. Instead, in early 2015, ministers from Egypt, Sudan, and Ethiopia met for a week in Khartoum, the capital of Sudan, where the Blue Nile merges with the other tributary, the White Nile. They emerged from secret negotiations to announce that they had agreed to a deal for managing the dam. The deal, they said, enshrined the principles that Ethiopia had the right to develop its economy by building the dam, while Egypt and Sudan had the right to insist that they did not suffer "significant harm" from the dam. Sudanese foreign minister Ali Karti called it "the beginning of a new page in relations between our three countries." Days later, Egypt's president, Abdel Fattah el-Sisi, made a first-ever speech to the Ethiopian parliament looking to "a future where all the classrooms in Ethiopia are lit and all the children of Egypt can drink from the River Nile."[4] A water war seemed to have been averted.

What remained unclear was how exactly Ethiopia could operate its dam without causing "significant harm" to its downstream neighbors. By 2017, with the dam approaching completion, there was still no actual deal on specifics. The parties had appointed French consultants to come up with suggestions for how such a deal might be concluded. Talking was better than fighting, but longtime Nile observers warned that the dispute might still have a long way to run.

It is hard to overstate the importance of the Nile to Egypt. The river is the only source of water for forty million farmers to irrigate their fields. It is not just farmers who depend on the river. Turbines in Egypt's own giant Nile barrier, the Aswan High Dam, constructed for them by Soviet engineers in the 1960s, generate electricity for Egyptian cities like Cairo, which has a population of almost ten million. Thanks to the beneficial terms of the old 1959 Nile Agreement, Egypt still has much more water per capita than most of its neighbors. Much of its irrigated farming, which takes 90 percent of the water, is hugely wasteful. So there is great potential to do things better. But existing water entitlements are a "red line" that the nation believes it cannot allow to be crossed. Not even by Ethiopia, which is the source of most of the river's water, and with a population now at ninety-five million might expect a share.

The stakes are almost as high for Ethiopia. The GERD will be the world's eighth-largest hydroelectric dam. It will flood some 650 square miles of forest and bush along the valley of the Blue Nile just inside Ethiopia's border with Sudan. The dam has huge political significance for the country and will at a stroke more than double Ethiopia's electricity-generating capacity. There are no loans from the World Bank or other outside investors. Instead, the country's citizens are being regaled with patriotic slogans to donate the $5 billion needed for the project. Civil servants devote at least a month of their wages every year to buying GERD bonds. The national lottery is pitching in too.[5]

"GERD is part of a larger social movement against poverty," I was told by Mulugeta Gebrehiwot, a former Ethiopian military and political leader who is now a fellow at the World Peace Foundation at Tufts University. "It has a symbolic impact in demystifying the fight against poverty—even a shoeshine boy can take part in building it by contributing coins and buying bonds that start at five US dollars."

Some of the national hopes riding on the dam have been exaggerated. An original promise to have the dam built by 2017 slipped. The design specifications may be flawed. The dam will have turbines capable of generating 6,000 megawatts of power—bigger, patriots boast, than Egypt's Aswan High Dam. Outside experts believe it will rarely have enough water to use all that generating power. The Nile is the world's longest river, but on the ranking of annual flow it comes in at ninety-first place. The Amazon has seventy-four times more water. One advisor to the Ethiopian government told me: "They will only be able to get 6,000 MW [megawatts] for 1 percent of the time; 3,000 MW would have been better."

The Ethiopian government insists that its dam is no threat to Egypt. That it is a hydroelectric dam, designed to catch water and pass it on downstream through turbines. That Ethiopia has no plans to divert water for irrigation. So Egypt will get its water, but on a different schedule. Those reassurances were reiterated by an international group of experts that convened in 2014 at the Massachusetts Institute of Technology to provide confidential advice for the then-secret negotiators.[6]

One of those experts—speaking on condition of anonymity—warned that Egypt did have two reasons for concern. The first is short-term. What happens while the reservoir behind the dam is being filled? The reservoir will have the capacity to hold back more than a year's flow of the Blue Nile. That represents two-thirds of the river's total flow into Egypt. So, while filling the reservoir for the first time, Ethiopia could, if it wished, turn the supply off for that year. Even filling gradually over five years would significantly impact flows into Egypt, especially if they were dry years.

The second concern for Egypt is that, by changing the natural flow of the river from a short monsoon burst into a year-round discharge through the turbines, the dam will allow Sudan to substantially increase the amount of water that it takes out of the river for irrigation. Currently, Sudan abstracts from the

Nile only about two-thirds of what it is entitled to under the 1959 agreement. Its only dam on the Blue Nile, the Roseires Dam, is too small to take more. But the Ethiopian dam will allow Sudan to abstract water throughout the year and divert it for year-round irrigation of crops. "The Sudanese government is already selling land leases for new farmland that will be irrigated when the GERD is completed," says Alex de Waal of the World Peace Foundation, who is an expert on Sudanese hydro-politics.

In a typical year, Sudan leaves around 5 million acre-feet of water in the river to which it is entitled. Egypt has come to rely on this Sudanese bonus. The MIT study says that if it took its full entitlement, it "could create a political crisis in the basin, if Egyptian farmers and policymakers blame their problem on Ethiopia and the GERD." As well they might.

The disputes between Egypt, Sudan, and Ethiopia over the Blue Nile are only part of the problem on the river. There are another eight nations along the river's banks, most of them along the White Nile. This smaller, though longer, tributary rises in the highlands of central Africa, before collecting in Lake Victoria and flowing north through South Sudan to Sudan, where it joins the Blue Nile at Khartoum.

Here too upstream nations are unhappy at Egypt's hydro hegemony. In 2010, four White Nile countries—Kenya, Uganda, Rwanda, and Tanzania—joined with Ethiopia to sign the Entebbe Agreement. They disowned the 1959 treaty and called for their rightful share of the river's waters. Burundi later also joined. Egypt and Sudan initially rejected the call, but more recently Egypt has been making overtures to the Entebbe signatories. Optimists claim to see signs that Egypt is in the mood for a wider peace deal over the Nile, if only to secure its downstream water rights once and for all.[7]

The good news is that this need not be a zero-sum game. You might imagine that, however diplomats try to dress things up, there is only so much water in the river. If upstream nations want more, for instance, then Egypt will have to make do with less. The truth is more complicated. And buried in the MIT report is the idea of reviving a century-old idea to manage the Nile in a way that would deliver more water for all. Even more encouragingly, it is something that could be fast-tracked by the completion of the GERD.

The idea is to manage the two biggest dams on the river, the GERD and the Aswan High, in tandem in a way that would reduce evaporation from reservoirs along the river.

As we saw in chapter 16, evaporation is a major source of water loss on the Nile, mainly from the Aswan High Dam in the Nubian Desert of southern Egypt. There are few worse places in the world to build a dam with a large reservoir than the Nubian Desert. The reservoir has a surface of some 2,000 square miles—make that sixty Manhattans. Subjected to almost year-round harsh sunlight, it loses up to 13 million acre-feet of water to evaporation in a typical year. This is almost a fifth of the average annual flow into the reservoir, and up to 40 percent in dry years.

The evaporation rates in the reservoir behind the GERD in the mountains of Ethiopia will be much less than farther downstream, because the climate is cooler and cloudier, and a reservoir in a steep valley will have a much smaller surface area. If the two countries could reach an agreement under which Lake Nasser was partly emptied and Ethiopia held water in the GERD on behalf of Egypt, then several million acre-feet of water could be saved.

There would be difficulties. Less water behind the Aswan High Dam would reduce its potential to generate hydroelectricity for Egypt. There is a quid pro quo here too, however, as MIT's Kenneth Strzepek told me. The GERD will double Ethiopia's electricity-generating capacity. For some time to come, the country will have spare power. That too could be part of the deal with Egypt. Water for power.

Of course, none of this makes allowance for climate change, which could drastically reduce Nile flows and upset all the calculations. But diplomats can only deal with what they know now. It would be a major concession for Egypt to hand over control of the river to another country—particularly its regional rival. But, just possibly, a grand settlement of the long dispute over who owns the Nile might create a solution that made hydrological as well as political sense for all. It is too early to talk about peace breaking out on the Nile. The new dam will create new options. Maybe, in the future, Egypt will not be the only "gift of the Nile."

24

IRAQ

Blood and Water

DURING THE RISE AND FALL of the Islamic State (ISIS), the terror group that for a while took over much of Syria and Iraq, one of the least discussed aspects was its battle for control of the country's large dams and water diversions—the vital infrastructure that has governed the fate of civilizations there since the days of ancient Mesopotamia. Behind the headline stories of brutal slaughter as the Sunni militants attempted to carve out a religious caliphate covering Iraq and Syria, blood was spilled to control the region's two great rivers, the Tigris and Euphrates. With engineers often fleeing as the forces of Islamic State advanced, the result could have been hydrological catastrophe—either deliberate or accidental. Many of the risks remain.

"Managing waterworks along the Tigris and Euphrates requires a highly specialized skill set, but there is no indication that the Islamic State possesses it," Russell Sticklor, a water researcher for the CGIAR, a global agricultural-research network, told me in 2014. The stakes were especially high after the Islamic State early that year captured the structurally unstable Mosul Dam on the Tigris. There have been other attempts to grab the dam—at the time of writing none of them successful. Without constant repair work, the Mosul Dam could collapse and send a wall of water downstream, potentially killing as many as a million people.

The Euphrates flows out of Turkey, through northeast Syria and along the length of Iraq, before reaching the Mesopotamian marshes, a region of reed beds and waterways where the river enters the Persian Gulf. The Tigris also begins in Turkey and flows farther east directly into Iraq, initially through territory controlled since the overthrow of Saddam Hussein by the Kurds. It follows a path parallel to that of the Euphrates before the two rivers mingle in the marshes. Together, the rivers water a region long known as the Fertile Crescent, which sustained ancient Mesopotamian civilizations for thousands of years. They were the first rivers anywhere known to have been used for large-scale irrigation, beginning about seventy-five hundred years ago. The first known water war was on the Euphrates, when the king of Umma in Sumer cut the banks of irrigation canals dug by his neighbor, the king of Girsu. Hydrologically, not so much has changed. In the Fertile Crescent, as much as in Palestine or down the Nile, water is worth fighting over at least as much as land. During recent decades, Iraq, Syria, and Turkey have come close to war over the two rivers. The main differences today are that the diversion dams are bigger and that they supply both hydroelectric power and water.

The Islamic State's quest for hydrological control of the territories it claimed for its caliphate began in northern Syria. In early 2013, it captured the old Russian-built Tabqa Dam, which barricades the Euphrates as it flows out of Turkey. The world's largest earth dam is a major source of water and electricity for five million people, including Syria's largest city, Aleppo. It also irrigates 400 square miles of farmland.

After ISIS captured the dam, its management was haphazard, to say the least. In May 2014, locals watched as the water level in the reservoir behind it, Lake Assad, dramatically fell. Engineers at the dam told the Arab news service Al Jazeera that their new masters had ordered them to maximize the supply of electricity. That meant emptying the reservoir through the dam's hydroelectric turbines, even though little new water was arriving from upstream. Seeing the reservoir running out of water, the dam's new masters took fright. They changed tack, cutting flows through the turbines and rationing electricity. That caused blackouts in Aleppo for sixteen to twenty hours a day.[1]

Three years later, Islamic State was in military retreat. American planes dropped bombs near the dam to help rebels trying to retake Raqqa, the closest city. There was panic that the dam could be breached, inundating Raqqa, where some one hundred thousand people were trapped by the fighting.[2]

From the Tabqa Dam, the Euphrates flows through Iraq. Near Baghdad, in the center of the country, its route is again blocked. The Fallujah Dam diverts water for massive irrigation projects that feed the country. ISIS captured the dam in April 2014 and stopped all flow downstream. This left towns such as Karbala and Najaf, a Shiite holy city, without water. But it also caused the reservoir behind the dam to overflow east, flooding farmland and homes as far as the outskirts of Baghdad. Later, the rebels abruptly reopened the dam, causing flooding downstream along the river.[3]

This mayhem may have had a purpose. Ariel Ahram, a security analyst at Virginia Tech University, suggested to me that the initial eastward flooding had been aimed at repelling Iraqi government forces attempting to retake the dam. Or it may have been a simple failure by ISIS fighters to understand the hydrology of the river and the consequences of how they operated the dams.[4]

Thankfully, ISIS never got control of the bigger Haditha Dam, between Tabqa and Ramadi. Iraq's second-largest dam regulates the river Euphrates for the whole of Iraq, providing water for irrigation and generating a third of the country's electricity. It keeps the lights on in Baghdad. In 2014, ISIS fighters repeatedly tried to capture it. If they had succeeded, said Sticklor, "it would have a potentially crippling effect on food production and economic activity in central and southern parts of the country." Releasing its waters would also have flooded large areas of southern Iraq. US troops knew this a decade previously. When they planned their invasion of Iraq from the south in 2003, they made the dam their first target, fearing that Saddam Hussein would release its waters across the desert to halt their advance. A decade later, employees at the dam said that Iraqi generals were prepared to open the floodgates against ISIS forces, rather than give up the dam.

On Iraq's other great river, the Tigris, the ISIS fighters made their prime target the Mosul Dam. They briefly took it in 2014. Mosul is Iraq's largest dam and the fourth-largest in the Middle East. It is 370 feet high and barricades the river just after it leaves Turkey, about 25 miles upstream of Mosul, Iraq's second-largest city. Since its completion in 1984, living downstream of the Mosul Dam has been risky—regardless of who is in control. The Mosul Dam is "the most dangerous dam in the world," a study by the US Army Corps of Engineers concluded in 2007. It still is.

The dam is built in a valley made of porous gypsum rocks. The water in its reservoir constantly dissolves the gypsum, creating sinkholes that threaten to undermine the dam itself. Iraqi engineers have been working around the clock for years, pouring tens of thousands of tons of cement into the holes beneath the dam to keep it from collapsing. We have to hope they continue to plug the leaks. The reservoir holds more than 9 million acre-feet of water. A 2004 study by the engineering firm Black and Veatch predicted a failure of the dam would send a flood wave 60 feet high rushing down the Tigris valley. It would flood Mosul city within three hours, inundating anywhere within 4 miles of the river.

Within three days, the flood would reach Baghdad, still about 12 feet high. The study estimated the death toll at more than a million people.[5]

A conference in Stockholm to discuss fears about the state of the dam in 2016 concluded that the "grave dangers threatening the population downstream" of the dam remained. Reinforcing the foundations needed to be done "as quickly as possible." Italian engineers moved in to do the work.[6] After six months, in May 2017, the Iraqi government claimed the dam was out of danger.[7] US engineers supervising the project said the job was only half-done.[8] In any event, the Stockholm conference concluded that the only solution was to either demolish the dam or build a new one downstream to be kept empty, ready to capture any future flood from the Mosul Dam.

The demise of ISIS as an occupying force at the end of 2017 much reduced the threat that the country's people face from dangerous dams and their sometimes even more dangerous operators. But the uncertain fate of the Mosul Dam makes clear that the long conflict has left a legacy of damaged infrastructure that could still hold some nasty surprises for anyone living downstream of one of the country's large dams.

25

MESOPOTAMIA

Who Killed the Garden of Eden?

THE MESOPOTAMIAN MARSHES that some call the Garden of Eden have been saved, even as chaos grows all around. Or have they? In October 2013, amid a wave of bombings on the streets of Baghdad and only weeks before the invasion of the north of the country by ISIS fighters, Iraq's Council of Ministers found time to approve the creation of the country's first national park. It was the centerpiece of a remarkable restoration of the vast wetland of reed beds and waterways where the rivers Tigris and Euphrates unite and flow into the Persian Gulf. That, at any rate, was the hope.

The Mesopotamian marshes have for millennia been the largest area of wetland in the Middle East. This strange enclosed world, twice the size of the Florida Everglades, sustained huge numbers of fish, large populations of birds, and rare animals such as the smooth-coated otter. The marshes are widely held to be the origin of the biblical story of the Garden of Eden, the paradise where Adam and Eve were created and from which they were subsequently expelled. But the wetland was never a wilderness. The marshes were managed. For at least five thousand years, they had been the home of the Ma'dan people, the "Marsh Arabs," who had made their waters their own, burning the reeds annually. Without the burning, ecologists say the waterways would have choked and the marsh would have dried up. By the mid-twentieth century, as many as half a million people grew rice, caught fish, hunted otters and birds, gathered reeds, and grazed their water buffalo in a world immortalized in Wilfred Thesiger's 1964 book *The Marsh Arabs.*[1]

Little changed in their watery world until the first Gulf War in 1992. In its aftermath, and apparently encouraged by the Americans, the Ma'dan rebelled against Saddam Hussein's rule in Baghdad, siding with their Shia fellows in neighboring Iran. Their rebellion failed, however, and Saddam took

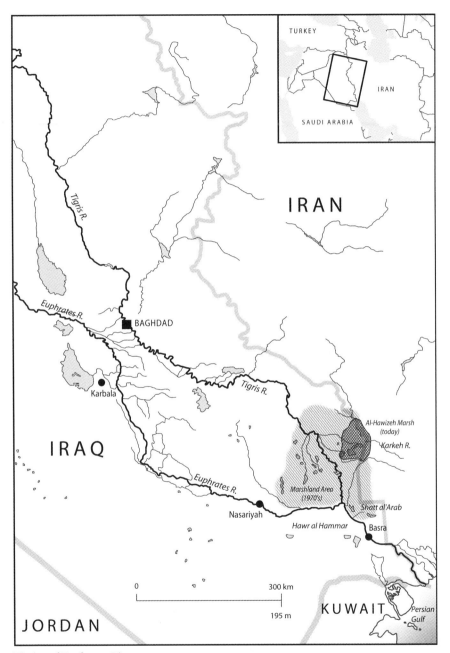

Tigris and Euphrates Rivers

his revenge by engaging in ecological warfare. In 1993, his military engineers dug huge drains to empty the marshes, diverted the two rivers away with huge canals, and raised embankments to prevent future flooding. By the end of the decade, more than 90 percent of the marshes had dried out.[2] The loss of this precious wetland was "a major ecological disaster, comparable to the deforestation of Amazonia," said Klaus Töpfer, then director of the UN Environment Programme (UNEP).[3]

Saddam always insisted that his vast engineering project was intended only to reclaim the marshes for irrigated farming to feed his people. It was not a self-evidently preposterous claim. Fifty years before, British engineers had proposed just such a project. As international anger about his draining of the marshes grew, I found on a shelf in the Institution of Civil Engineers in London a report called *Control of the Rivers of Iraq*. It was published in 1951 by the Iraqi Irrigation Development Commission and written by a senior British colonial engineer of the day, Frank Haigh. In one chapter, Haigh proposed diverting the Euphrates into a "Third River" that would reach all the way from Baghdad to the Gulf, bypassing the marshes. He suggested installing sluices to restrict the flow of the Tigris into the marshes. His purpose was to capture the water then "wasted" in the marshes and use it for irrigating crops on the drained land.

It was evident that Saddam's scheme was essentially a carbon copy. In fact, construction of part of the Third River, which he continued, had begun under British supervision in the 1960s. "We did the early design work for the Third River," Bill Pemberton, at the British firm of consulting engineers Mott Mac-Donald, told me as Saddam's work progressed. "We built our bit, about 15 miles. The rest they now seem to have done themselves."

Whatever Saddam's long-term aims, and whatever blueprints he drew on, most agree that his immediate political purpose in draining the marshes was to empty them of the rebellious Ma'dan. In that, he succeeded. A marsh population of half a million was reduced to fewer than eighty thousand. Refugees scattered across the Middle East as their homeland was turned into a salt-encrusted desert, dotted with burned-out villages and poisoned with the pesticides that Saddam's soldiers poured into the marshes to kill any remaining life.

There things remained until the Second Gulf War in 2003. We all watched TV images of American forces crossing the desert of southern Iraq, intent on overthrowing Saddam. Few realized that, until a decade before, most of the land they crossed had been covered in lakes, reed beds, and waterways. Just as Saddam's overthrow brought hopes of political liberation in Iraq, so it brought hopes among the Ma'dan that they could return home and among environmentalists that the marshes could be re-created. And so it happened. The Ma'dan tore down many of Saddam's barricades to let the water from the rivers back. Their makeshift engineering was imperfect, but some areas recovered, while the interim Iraqi government in Baghdad soon declared that it wanted international funding to help it restore the rest.

Baghdad's calls found vocal outside assistance. Foremost among them were Azzam Alwash, an Iraqi exile living in California, and his American wife, a wetlands ecologist. Before the rise of Saddam, Azzam's father had been an irrigation engineer, helping outsiders grow rice. As a boy, Azzam had joined his father duck-shooting in the marshes at weekends. With the rise of Saddam, they fled the country, but now Azzam wanted to go back and mastermind reflooding the marshes.[4] He and his wife, Suzie, gathered scientists from around the world and, in the heady days after Saddam's fall, gained the support of the US State Department, UNEP, and the World Bank's Global Environment Facility. "On our advice, the government is constructing new banks and canals to keep parts of the marshes wet all year round and help sustain the local economy," he told me at the time. "We hope the government will very soon turn the marshes into a national park." In 2013, it did.

Optimism was high that year. Peace and stability seemed finally to be returning to Iraq. A year of exceptional rains had left the marshes full of water. Conservationists reported that all 278 bird species ever recorded in the marshes remained, including the endemic Basra reed warbler and Iraq babbler. "They had hung on in small spots. When the water spread again, so did the birds," said Richard Porter of Birdlife International. "It shows how resilient nature can be and gives hope that other lost wetlands can be restored."

The Ma'dan were not, of course, going back to the old ways, Azzam Alwash agreed when we spoke. "Marsh Arabs used to live isolated lives out among the reed beds. Today, they mostly live in towns on the banks. They have cell phones and satellite dishes. They damage the fisheries with electric fishing. They shoot the birds. We see garbage floating in the marshes," he told me. "They have to live, but our task will be to help them use the resources more sustainably." In the newly created national park, he said, "we hope tourists will provide another source of income and an incentive to protect the marshes. We must continue to manage them like a garden." Like the Garden of Eden.

But by 2015, his hopes had been dashed. Water levels in the marshes had fallen to their lowest levels since Saddam's engineers invaded in the 1990s. The problem was that, during the long years of conflict, much had changed upstream on the two rivers that once supplied its waters. While Saddam's barriers were now removed, much less water was coming down the rivers than twenty years before. In the marsh town of Al Chabaish, "the amount of water from the Euphrates reaching the marshes is less than a tenth of what is needed," Suzie Alwash told me. "Water buffalo can no longer drink from the marshes, and fish [deaths] are common."

Who was to blame for the dried up Euphrates? The Iraqi parliament pointed the finger at Turkey, where hydroelectric dams now blocked the river's upper reaches. The giant Ataturk Dam had cut flows into Iraq by a fifth. Perhaps even more damaging, the dams had eliminated the peak spring floods, which had inundated the widest areas of marshes. Iraq accused Turkey of breaching a deal to deliver a minimum of 0.4 acre-feet of water per second down the Euphrates.[5]

There were other potential hydrological villains. Iraqi officials also blamed ISIS, which, as we saw in the last chapter, had disrupted flows after capturing dams.[6] Iran may have been to blame as well. Twenty years before, up to a third of the flow down the Tigris into the marshes had come out of Iran down the river Diyala, called the Sirwean in Iran. But since then, the Iranians had built several diversions that removed half this flow for irrigating fields across southwest Iran.[7] It also captured much of the flow of the river Karkeh, which once drained directly into the marshes.[8]

The final culprit may have been the Iraqi government itself. In 2015, anxious to exploit the river's waters in a hearts-and-minds battle to hold back ISIS, it was diverting most of the water approaching the marshes down the Euphrates at the holy city of Najaf to irrigate the area's rice fields for Shi'ite farmers, who backed the government. This was contrary to government policy to "not allow rice growth in years of little water," Azzam Alwash told me.

"With more dams being built all the time, the whole region is in a water crisis," said Hassan Partow at the UN Environment Programme. For more than a decade, UNEP has been trying to broker a deal among the nations on the two rivers that would allow them to take water while engineering a spring flood-pulse into the marshes. But the signs are not good. The threats to the marshes today seem as great, if not greater, than in the time of Saddam.

26

TIBET

Asia's Water Tower

THE TIBETAN PLATEAU is one of the largest sources of freshwater on an increasingly thirsty planet. It is Asia's water tower. Meltwaters from its vast ice sheets and runoff from the monsoon rains supply some two billion people across East and Southeast Asia with water for irrigation and drinking, and offer the promise of unparalleled hydropower. The question is: who owns this water?

China, as the upstream state where the rivers rise, holds most of the cards. As easily the world's largest builder of dams, it is increasingly playing those cards. Altogether, it shares 111 rivers and lakes with seventeen neighboring countries and is upstream on almost all of them, says Oxford University's David Grey. But as China takes ever more of the water, what are the rights of its downstream neighbors? How complete is China's hydro hegemony? And what happens if one day Chinese dams stop the waters of great international rivers from flowing? Questions like these are fast becoming incendiary geopolitics.

Until recently, China mostly dammed rivers that flowed only within its borders, such as the Yellow River and the Yangtze. To meet soaring demand for energy and irrigation, its engineers have begun eying each of the five great rivers that reach the ocean through neighboring countries. An earlier chapter looked at one of them—the Mekong. But the other four are equally contentious: the Brahmaputra, Irrawaddy, Salween, and Indus. Concern about China's intentions is growing because of its adherence to a doctrine it offered at the UN in 1997. In explaining why China was one of only three countries that refused to sign a treaty aimed at encouraging international agreements on cross-border rivers, it claimed "indisputable territorial sovereignty over those parts of international watercourses that flow through its territory." No matter what happens downstream, the water was theirs.

This is troubling, to say the least. While China may swat away challenges from its smaller neighbors, like those downstream on the Mekong, what about India, a fellow Asian superpower? India already has chronic water shortages and believes it can ill afford to countenance current Chinese plans for the Brahmaputra. The 2,000-mile-long river, known in China as the Yarlung Zangbo, was until recently one of the planet's last great untamed rivers. But after building a few small dams on its tributaries, Chinese engineers in 2015 completed the first barrier on the river's main stem, the $1 billion Zangmu Dam, near the country's borders with Bhutan and India.[1] They have much bigger plans, especially in the Zangpo Canyon, the deepest canyon in the world and slightly longer than the Grand Canyon in the United States.

The Zangpo Canyon formed millions of years ago as the river carved its way through the eastern Himalayas. Engineers say it is probably the best remaining untapped hydroelectric dam site anywhere in the world. China would like to construct two giant hydroelectric plants in the canyon. Each could deliver roughly twice as much power as the Three Gorges Dam on the river Yangtze. The Motuo Dam would deliver 38,000 megawatts, and the Daduqia 42,000 megawatts.

These plans are making Bangladesh and India very nervous. India sees the Brahmaputra as its largest untapped water source. It has plans to divert some of its flow to distant southern India. The real victims, however, could be the 160 million inhabitants of Bangladesh, who rely on the river for irrigating their crops during the long dry season. The competing projects could lead to a resource conflict between India and China, and an environmental catastrophe for Bangladesh, according to Robert Wirsing of Georgetown University's School of Foreign Service in Qatar.[2]

Other neighbors are growing restive about China's water plans. Notably its former staunch ally, Myanmar (formerly Burma). In 2009, the Burmese generals gave China permission to start building the Myitsone Dam on the Irrawaddy inside northern Myanmar, even though 90 percent of the electricity from the 6,000-megawatt plant would go to China. After protests from the local Kachin people, during which dozens of people were killed, reformist president Thein Sein suspended construction in 2011. A total of six other Chinese dam projects in Burmese territory on the Irrawaddy, and six more on the Salween, were approved by the generals but are now in cold storage. Many, including the Myitsone Dam, would flood remote and rare forest ecosystems. Meanwhile, Myanmar has its own plans. Its proposed 6,000-megawatt Mong

Ton Dam on the Salween near the Thai border could become the largest ever dam project in southeast Asia.[3]

Farther west, on the Indus in northern Pakistan, China is building the Karot Dam and has signed an agreement for the 7,000-megawatt Bunji Dam. The latter plan has angered India, which claims the territory. Local Pakistanis also have reason to be fearful. The dam is close to the epicenter of an earthquake that killed more than one hundred thousand people in 2005.[4]

In the future there could be many more such acrimonious disputes, and not just in Asia. A quick look at the atlas shows the extent of the problem. The world is full of river basins shared between countries. More than 40 percent of the world's people, in 148 countries, live in 263 river basins that straddle international borders. Those rivers account for 60 percent of global river flow. A host of upstream nations threaten to grab the water before equal numbers of downstream nations can fill their cups. The Danube drains nineteen countries; the Nile passes through eleven; the Rhine, Niger, and Congo through nine; and the Zambezi through eight. Two-thirds of these transboundary river basins have no treaties for sharing their water. Several, including the Euphrates, Tigris, Jordan, Senegal, and Indus, pass through current or recent war zones. Each is the scene for a potential water war.[5]

What treaties there are often date back to colonial times. The Nile is theoretically governed by deals drawn up by the British in 1929 and 1959, which allocate all the water to downstream Egypt and Sudan and none to the nine upstream nations. Other river treaties are extremely weak, like the legal regimes of the Lake Chad Basin Commission or the Mekong River Commission. Or they are restricted to narrow data-sharing or technocratic remits, such as the OMVS, which is devoted largely to hydroelectric developments on the river Senegal, while largely ignoring their downstream impacts.[6]

Africa—a continent of haphazard national boundaries created in the days of imperial rule and maintained today because anything else would bring chaos—is full of countries dependent on their neighbors for most of their water. It has eighty transboundary rivers. The isolationist Islamic republic of Mauritania, in West Africa, gets around 90 percent of its water from the Senegal River coming out of Guinea and Mali. Tiny Gambia is similarly dependent on Senegal and Guinea. Botswana relies for 94 percent of its water on the Limpopo out of South Africa, and the Okavango, which rises in Angola and passes through Namibia before draining into the Okavango wetland, where tourists contribute a tenth of Botswana's GDP.

Worldwide, at least twenty nations get more than half of their water from their neighbors. Many should by rights get more than they do. Mexico receives virtually none of the flow of the Colorado River out of the United States. Little of the water in the river Jordan reaches the country named after it. Ethiopia is snuffing out the Omo and with it Kenya's Lake Turkana. In Central Asia, the river Ili has shrunk by two-thirds by the time it leaves China for Kazakhstan. The Karkeh, flowing west out of Iran, rarely makes it to Iraq anymore. Iran, meanwhile, has rarely seen the Helmand flow west out of Afghanistan since the 1990s. Given the extent of the damage caused by such disruption to natural river flows, it is perhaps a wonder that there have not been more out-and-out wars.

The international community has been slow to wake to this issue. In 1997, the UN adopted the Watercourses Convention. But it did not lay down hard and fast rules for sharing waters. It merely asked countries to sign up to a statement of principle that nations should ensure the "sustainable and equitable use of shared rivers." Even so, three countries voted against. They were China; Turkey, which is busy dam- building in the headwaters of the Tigris and Euphrates; and Burundi, which is at the head of both the Nile and the Congo. Others with contentious hydrological geopolitics such as Israel abstained. To come into force, the treaty required thirty-five nations to ratify it in their legislatures. That took until 2014 to achieve. Many major nations that signed the deal at the UN have so far refused to ratify, include the US and Britain, which was an original sponsor of the treaty.

This is dangerous complacency. Dam construction generates international tension when the rivers they block do not have functioning water-sharing agreements. Ethiopia's work on the Grand Ethiopian Renaissance Dam on the Nile, upstream of Sudan and Egypt, is an obvious example. There are many others. Guinea threatens to barricade the river Niger, which could dry out Mali's Inner Niger Delta, a wetland jewel on the edge of the Sahara. In 2012, Vladimir Putin visited the mountain states of Tajikistan and Kyrgyzstan in Central Asia, where he promised money for more dams on the Amu Darya, once known as the Oxus, and Syr Darya Rivers to generate hydropower in those countries. He ignored opposition from downstream Uzbekistan and Kazakhstan, who fear the dams will deprive them of summer flows to irrigate their cotton crops. Mozambique fears Zambian plans on the Zambezi, and so on.

≈≈≈

Having a treaty may not on its own resolve deep-seated water disputes, but it can provide a forum for resolving them. That has worked so far in the acrimonious disagreements between nuclear powers India and Pakistan over the river Indus.

Southern Doda in Indian Kashmir is bandit country. Pakistan claims the region, and Muslim youths regularly cross from India into Pakistan for training, returning as guerrillas to fight Indian security forces. The conflict has been going on for decades. It usually hits the headlines only when foolhardy Western tourists get caught up in it. Some people believe that the world's first nuclear exchange could one day be triggered between India and Pakistan because of events in the foothills of the Himalayas. If so, it may not be over guerrillas or terrorists or even border incursions by regular soldiers. It could be over water. For Kashmir is the gateway through which Pakistan receives most of its water along tributaries of the Indus River.

Pakistan's 190 million people would be in trouble without the Indus. It runs the length of their country. Its waters irrigate most of the country's crops and generate half its electricity. But the Kashmir gateway is an Achilles' heel. India and Pakistan have been in armed conflict three times since the two states were formed on British withdrawal from the subcontinent in 1947, and the first conflict arose when India intervened in Kashmir to cut the flow of tributaries of the Indus on which Pakistan relied. After a decade of negotiations following this conflict, the World Bank brokered the 1960 Indus Waters Treaty between the two countries. It bound them to share the river's flow, with each taking water from three tributaries. The Chenab River was given to Pakistan and has ever since been the biggest source of water for Punjab, the breadbasket of Pakistan.

Geography dictates that India always has the potential to stop Pakistan's water. Forty years after the treaty was signed, India began building a dam on the Chenab just before the river crosses the border into Pakistan. The 525-foot-high, billion-dollar Baglihar Barrage was completed in 2008. Pakistan said it was a clear breach of the treaty. India denied any breach. It said the dam was intended only to generate hydroelectricity. India would not remove any water from the river, only hold back water to discharge downstream through turbines. But Pakistan said that the dam would allow India to hold it to ransom in any future crisis between the two nations. It could use the barrage to starve Pakistan of water during the critical winter planting season. It could cause famine.

The two sides went to arbitration, and the judgment broadly found in India's favor. Even so, tensions over the dam have not gone away. Pakistan is demanding that India give advance notice of releases from the dam to help its management of floods downstream.[7] Bickering over the treaty has become a diplomatic norm.[8] India ups the ante with plans for three new Indian hydroprojects—the Kishanganga, Ratle, and Miyar Dams—on tributaries whose waters are allocated to Pakistan.[9]

The stakes are huge. For almost four decades the Indus Waters Treaty has been held up as a model for other countries with water disputes. In truth, though, it has been enforced largely at the insistence of the World Bank, which refused to give either country loans for new dams unless they maintain its provisions. With other major international dam builders and funders, notably the Chinese, now busy on the river, that system of enforcement no longer has the power it once did.

VIII

When the rivers run dry . . .
civilizations fall

27

ELISHA'S SPRING AND
THE RIDDLE OF ANGKOR

THE FIRST KNOWN permanent human settlement was on the west bank of the Jordan River. Jericho was constructed some nine thousand years ago as the Ice Age receded and *Homo sapiens* began to prosper in the warmer postglacial world. The settlement was modest enough. It covered barely 10 acres and had a thick defensive wall. Inside were a few hundred people and a stone tower, which survives today as reputedly the world's oldest human-formed structure. Close by was a spring, recorded in the Bible as Elisha's Spring. It was the reason for Jericho's existence. As the Book of Kings puts it, "The hand of the Lord came upon him. And he said, Make this valley full of ditches. For thus saith the Lord, Ye shall not see wind; neither shall ye see rain, yet that valley shall be filled with water." The spring gushed into ditches that distributed its water to the fields and orchards.

The original town was destroyed by floods in the valley about eight thousand years ago. Jericho has been rebuilt and lost several times since, but the spring and the farming that it sustained just kept going. The area around was vital to the development of modern farming. It was among the first places where people cultivated wild grains. By six thousand years ago, its inhabitants were growing peas, beans, olives, vines, and figs. Today the spring still delivers water, at a regular 20 gallons a second, into a small pool. The farms still grow crops in what must be one of the longest-lasting and most durable agricultural systems in the world.[1]

During the thousands of years since the fields of Jericho first flourished, many grander civilizations have come and gone in the Middle East, many of them based on apparently more sophisticated methods of catching and manipulating water. These are the places where Western civilization is deemed to have begun. It was around seventy-five hundred years ago that Sumerians in the Fertile Crescent constructed the first large irrigation systems using river water. They diverted water from the Tigris and Euphrates Rivers down long canals and erected earth defenses against the spring floods. Building on this agricultural prosperity, they began to construct great cities too. Cities like Ur, Kish, and Uruk, where the first writing was produced and the first sciences developed. Uruk eventually had a population of fifty thousand.[2]

These early Sumerian cities fought the first water wars. Some five thousand years ago, the king of Umma cut the banks on canals that neighboring Girsu had dug to the Euphrates. The breaches spilled water across the plain and destroyed Girsu's ability to feed itself. But the kings of Girsu were not beaten. They dug a new canal to capture the waters of the Euphrates' twin river, the Tigris. That gave them a larger empire than their rivals, and they ultimately saw the end of Umma.

The Sumerian fields gradually became blighted. The bumper wheat harvests began to fail. Wheat gave way to barley. Then the barley too waned. With that, the fields became barren, the civilization foundered, and the land returned to desert. When the British archaeologist Sir Leonard Woolley excavated Ur in the 1930s, he wondered at the contrast between the great civilization he was uncovering and the barren land around him: "To those who have seen the Mesopotamian desert, the evocation of the ancient world seems well-nigh incredible. . . . Why, if Ur was an empire's capital, if Sumer was once a vast granary, has the population dwindled to nothing, the very soil lost its virtue?"[3]

Woolley's successors believe they have solved the riddle. The problem seems to have been tiny amounts of salt coming down the river and accumulating over the centuries in the fields, eventually poisoning soils and crops. Cuneiform tablets of thirty-eight hundred years ago describe a farm system in its death throes, with "black fields becoming white" and "plants choked with salt." That would explain why wheat, which is less tolerant of salt than barley, gave way first. Manipulation of the rivers made the Sumerian cities, but ultimately it destroyed them too. Salt chased civilization through Mesopotamia as mercilessly as any barbarian horde.[4]

While the fields of Sumer turned to dust, new cities grew up farther north, around modern Baghdad. These societies constructed canals to carry water from the Tigris to irrigate fields for 200 miles on either side of the river. This remaking of the landscape far outstripped the efforts of the Sumerians. Indeed, nothing bigger has been built in this region until today. It was "a whole new conception of irrigation, which undertook bodily to reshape the physical environment," said the American archaeologist Thorkild Jacobsen.[5] In places, the canals were 300 feet wide and, according to Sir William Willcocks, a leading Victorian engineer, "must have been capable of quite crippling the Tigris."[6]

The Persians, who by then ruled the area, learned to keep salt at bay by avoiding overirrigation and planting weeds during the fallow season to keep the water table low. But they faced a new problem when the canals washed river silt into the irrigation systems. They solved it by employing thousands of slaves to dredge the waterways. When Islamic rulers took over from the Persians, the dredging was left undone, and at some time in the twelfth century silt overwhelmed the system. Willcocks, who spent several years dreaming of reconstructing the canals, wrote of a "terrible catastrophe which in a few months turned one of the most populous regions of the earth into desert."

The lessons for modern times in the threats posed by salt and silt are clear. For historians, however, there is something else here that is, on the face of it, a little strange. Why cities? It is not obvious that rich agrarian societies should need them at all, and yet from Mesopotamia to the Nile in Egypt, and from the Indus in Pakistan to the Yellow River in China, a series of great civilizations grew up by irrigating arid areas from mighty rivers—and they all had cities.

One school of thought holds that cities were largely born because these agrarian societies needed new kinds of social organizations to collect, distribute, and contain water on a large scale. They had to hire farmers or coerce slaves into digging and maintaining dikes and canals and watching for floods; and they needed to develop scientific skills like astronomy and mathematics to predict nature's whims. It was the American historian Karl Wittfogel who coined the phrase "hydraulic civilizations" to describe societies that are organized primarily around the need to manage water. He argued that they often required cities to do this. "It is the combination of hydraulic agriculture, a hydraulic government, and a single-centered society that constitutes the institutional essence of hydraulic civilization," he wrote.[7]

Not all academics agree. But it remains remarkable how many great early civilizations emerged in environments where management of water was the first priority. Wittfogel contended that until the industrial revolution, the majority of human beings lived within the orbit of hydraulic civilizations. Ancient Egypt developed its water-management skills so well that it maintained population densities along the Nile that were double those of modern France. In the Americas at one time, three-quarters of the population lived in a few small centers of hydraulic civilization in Mexico and Peru.

Much of Europe did not need such sophisticated water management, because it could rely on rainfall to irrigate crops. Nonetheless, it was those arid areas that did have to manage water on a large scale that innovated and developed in other ways too. In the Middle Ages, Córdoba, the capital of Moorish Spain, sustained a population of more than a million people through irrigated agriculture, at a time when the largest city north of the Alps was London, with a mere thirty-five thousand people.

It was not just desert civilizations that depended on sophisticated management of water. One of the most stupendous preindustrial civilizations was centered on Angkor Wat in Cambodia. The Khmer civilization rose around AD 800, reached its height under Hindu kings in the twelfth and thirteenth centuries, and crashed in the early fifteenth century. Its influence spread throughout Southeast Asia, and its great capital, one of the true wonders of the world, was set in the jungle at the head of the Great Lake on the Tonle Sap, the reversible tributary of the Mekong. At one time it was probably the largest urban area on the planet.

Angkor is known today for its array of temples. The thousands of carvings depicting fish and fishing that festoon the walls of the temples are testimony to the importance of the lake to the wealth of the empire, and to the enduring productivity of the ecosystem. But recent research has revealed that the Khmer were about much more than temples. They, too, were remaking their landscape, and water was at the center of their works. The temples that we see today were simply the ceremonial heart of a huge suburban landscape that, at its height, was "by far the most extensive pre-industrial city in the world," says Roland Fletcher, an Australian archaeologist who has worked there for many years. "It was like modern-day Los Angeles."[8]

Satellite pictures show what is not so visible from the ground: huge networks of canals spreading out from the temples, linking reservoirs and rice paddies, ports and suburbs. Homing in on areas identified from the air, archaeologists are digging up the remains of communities engaged in mining and boatbuilding, weaving and manufacturing salt. All this, it is increasingly clear, was dependent on sophisticated water management.

Visitors to the Angkor temples can still see huge moats and reservoirs dotted among the temples. The largest reservoir, the West Baray, is 5 miles across and still holds water. Locals go swimming and take boats out to a small temple on an island in the middle. Other reservoirs, such as the East Baray, are now empty and encroached on by jungle. Yet, as huge as they were, these barays were only baubles. The new evidence is that the workaday job of keeping the hundreds of thousands of Angkor inhabitants and their fields watered was done by waterways connected to suburban pools and areas of rice paddy that grew three and sometimes four crops a year. It was this that underpinned the wealth and the splendor of the rulers and their temples.

The Khmer civilization, arguably the greatest that Southeast Asia has ever seen, eventually faltered. What went wrong? The latest evidence points to an environmental catastrophe brought on by a failure of the water supply. Matti Kummu, a Finnish hydrologist who has interpreted the hydrology for an international team of archaeologists working in Angkor, says that the barays and ponds and canals and paddy fields all drew their water from streams draining from the surrounding hills into the Great Lake. The great urban center was established beside the lake precisely to use those waters.[9]

The channels that collected the water and delivered it to the suburban complex were beautifully built and sometimes as much as 130 feet wide. But they were also artificial channels and prone to break out of their allotted course. They required constant maintenance. Kummu says that over the centuries, one of them, the Siem Reap Channel, started to take more and more water, and as it did, it began to cut its bed ever lower. In the end, the channel became so low that it could no longer deliver water to the barays and the ponds. The distribution network was left high and dry. (The West Baray has water in it today, Kummu points out, only because it gets it from another source.)[10]

I found the Siem Reap Channel in among the temples. It is still flowing and has indeed eroded a deep gorge some way from its original course, where several piers of an old stone bridge stand forlornly in the bush. Locals have rather elegantly installed a large waterwheel that scoops water out of the bottom of the channel and pours it into a pipe at the top to irrigate a small tree nursery. Farther on, where the channel skirts the East Baray, it is 40 feet below the surrounding land and evidently could not fill the reservoir.

There are many unanswered questions about the collapse of the Khmer civilization. But it is increasingly clear that just as management of water created this quite mind-boggling civilization in the jungle, so too failures of management must have destroyed it. Maybe that is also what happened in the jungles of Central America, where the ruins of the Mayan civilization, like those of Angkor, today emerge from regrown rainforest. Here, great pyramids loom amid the trees as a reminder of a culture that, starting some three thousand years ago, cleared large areas of the forest and converted it into fields and cities and suburban areas that became one of the most densely populated regions on the planet. The Mayans created an urban society that established universities for mathematics and astronomy, pioneered the cultivation of corn, and built large reservoirs, viaducts, and canals to grow crops all year round.

Reservoirs at Tikal, a Mayan city in northern Guatemala, could hold enough water to serve ten thousand people for up to eighteen months. At Coba, the Mayans built dikes to increase the capacity of a natural lake. The central Petén region, the civilization's jungle heartland, may have had a population of more than ten million. After two millennia of success, however, the Mayan civilization suddenly and mysteriously collapsed in the ninth century. The collapse was so complete that when the Spanish arrived in the region a few hundred years later, just thirty thousand people, less than 1 percent of the former population, were left roaming the jungle. Moreover, life for them was so poor that even expert looters and pillagers like Hernán Cortés and his army almost died of starvation there.

What went wrong? The most likely theory is that changing climate overtook the Mayans. Tree rings and lake cores reveal three catastrophic droughts in the final century of the civilization, which peaked in 810, 860, and 910. The lakes all but dried out. Presumably the reservoirs suffered similarly. Perhaps there were brutal wars for what water remained. Perhaps disease and starvation overtook the parched cities. At any rate, a civilization that prospered for two millennia by taking advantage of the fecundity of the rainforest and its water supplies eventually succumbed when the climate turned against it and the rivers and reservoirs dried up.[11]

What can we make of all these tales of collapse? Two things. First, that civilizations built on intensive use of water often find themselves highly vulnerable. Either to climate change or to insidious and destructive elements in their own systems of water exploitation, like salt and silt or the eroding power of artificial water channels. Second, that less intensive and less grand uses of water—such as those employed around Jericho—can be more flexible in the face of change and so can be longer lasting. Jericho never grew as big as the famous hydraulic civilizations around the Middle East. But while those civilizations fell long ago, farmers still make a living in Jericho. That too may offer lessons for our management of water today.

28

ARAL SEA

The End of the World

About 3 miles out to sea, I spotted a fox. It wasn't swimming. The sea as marked on the map was no longer a sea. The fox was jogging through endless tamarisk on the bed of what was once the world's fourth-largest inland body of water. In the past fifty years, most of the Aral Sea in Central Asia has turned into a huge uncharted desert. No human has ever set foot on most of it. But wildlife is moving in. It cannot be long before someone decides that it should be protected as a unique, virgin desert. At present, though, such is the scale of what has happened here that the UN calls the disappearance of the Aral Sea the "greatest environmental disaster of the twentieth century."[1]

Until the 1960s, the Aral Sea covered an area the size of Belgium and the Netherlands combined and contained around a billion acre-feet of water. It was renowned in the Soviet Union for its blue waters, plentiful fish, stunning beaches, and bustling fishing ports. Many atlases still show a single chunk of blue. The new reality is very different, however. The sea is broken into two hypersaline pools containing less than a tenth as much water as before. The beach resorts and promenades where Moscow's elite once spent their summers now lie abandoned. The fish disappeared long ago. As the fox and I peered north from near the former southern port of Muynak in Uzbekistan, there was no sea for 100 miles.

What has caused this? The answer lies in the death of the two great rivers that once drained rainfall from a huge swath of Central Asia into the Aral Sea. The biggest is the Amu Darya. Once named the Oxus, it was as big as the Nile. In the fourth century, Alexander the Great fought battles along its waters as he headed for Samarkand, creating the world's largest military empire. The river still rushes out of the Hindu Kush in Afghanistan. But like its smaller twin, the Syr Darya, from the Tian Shan Mountains, it is today largely lost in the desert lands between the mountains and the sea.

During the twentieth century, these two rivers were part of the Soviet Union. Soviet engineers diverted almost all their flow—around 90 million acre-feet a year—to irrigate cotton fields in the desert. Perhaps nowhere else on earth shows so vividly what can happen when rivers run dry. On my journey to the region, I found a landscape of poison, disease, and death in what was once one of the most prized areas of the Soviet empire. I found mismanagement of water on an almost unimaginable scale. More disturbing still, I found that in the aftermath of the collapse of the Soviet Union, nobody seems to have the vision or the will to rethink how this land and its rivers might better serve the people living here. Promises made to the international community to bring back the sea are rhetoric as empty as the sea itself. Even less water is reaching the Aral Sea today than in Soviet times.

In fact, most years now, no water at all makes it all the way down the Amu Darya. The Syr Darya, meanwhile, has less than a quarter of its former flow, most of which is corralled into a small fishing reservoir in its former delta, rather ambitiously called the Little Aral Sea.[2]

In August 2014, satellite images revealed that for the first time the main surviving portion of the Aral Sea proper, in its eastern basin, had completely dried up. It was probably the first time this had happened for six hundred years, when the Amu Darya became for a while diverted into the Caspian Sea.[3] A smidgeon of water returned to the basin with heavier rains the following year. But from now on the Aral Sea, once the world's fourth-largest inland sea, is little more than a seasonal desert salt lake.

Central Asia has a long tradition of using the waters of its two great rivers to grow crops. Much of the region has long been covered in orchards, vineyards, and grain fields. Russian czars in the nineteenth century first saw that the combination of near-constant summer sun and water from the great rivers could produce cotton harvests to rival those of the United States. But it was the Bolsheviks who got down to business. Lenin lectured the southern republics of the new Soviet Union in 1921 that "irrigation will do more than anything else to revive the area, bury the past, and make the transition to socialism more certain." Under his successor, Stalin, the region's farms were turned into Moscow-run collectives, growing cotton for the textile mills of European Russia. "Commissar Cotton" had arrived. An ever-growing network of irrigation canals supplied water to billions of cotton bushes planted each spring on millions of acres of fields. Nations of nomads, cowboys, and orchard tenders were turned into a near-slave society of cotton pickers.

By 1960, the canals were removing a staggering 32 million acre-feet of water from the rivers. The Aral Sea remained full, partly because the rains had been good and partly because the irrigation systems eventually returned much of the water to the rivers as drainage. Then, Moscow demanded more. Between 1965 and 1980, the area of irrigated land more than doubled, and water abstraction tripled. Central Asia became one of the largest irrigated areas on the planet, with some 20 million acres of fields crisscrossed by canals that could stretch to the moon three times over. By its own lights, all this was a dramatic Soviet success story: central planning at its finest, with everyone yoked to the common purpose of clothing the Soviet empire.

By the 1980s, 85 percent of all the fields in the Aral Sea basin were growing cotton. Orchards and vineyards, vegetable patches and wheat fields, even sports fields were given up. Almost every citizen was drafted to pick cotton from the searing summer, when temperatures could reach 120 degrees Fahrenheit, until November, when frost froze the fingers. Prisons, mental asylums, and schools were emptied, factories and offices were shut, and herds of animals were abandoned for the duration. Nobody was excused: not nursing mothers, not students, not doctors, not their patients. Only government officials remained at their desks, counting cotton.

But this marvel of organization carried the seeds of its own destruction. The newest canals were delivering water to the driest areas with the poorest soils. Farmers poured over 6 feet of water onto these fields each year, whereas the earlier fields required just over 3 feet. Increasing amounts of water never returned to the river in drainage. Instead, it accumulated in waterlogged soils, evaporated from fields, or drained into the desert, where new lakes formed.

Meanwhile, and most devastatingly for the Aral Sea, in the early 1960s, engineers dug the Karakum Canal, which took much of the flow of the Amu Darya west for some 800 miles to Turkmenistan, the driest, emptiest, least populated republic in the Soviet Union. The canal was the longest and biggest irrigation canal in the world, and it turned Turkmenistan into the USSR's biggest and most profligate water user. It still is. In its first fifty years, the canal has taken nearly 500 million acre-feet of water right out of the basin of the Aral Sea. None of it ever returns, even as drainage water. It was after the completion of the Karakum Canal that the sea began to empty in earnest.

Many have called the emptying of the Aral Sea a Soviet blunder.[4] The truth is more chilling. The draining of the sea was in fact entirely deliberate. In a museum at Nukus on the Amu Darya delta, I found a series of maps drawn by Soviet engineers in the 1970s depicting the planned demise of the sea. By 2000, they expected it to be almost empty. Their plans once decreed that the seabed itself should be converted to cotton production as it dried up.

Today Commissar Cotton is gone, but his legacy remains. While a market economy has emerged for crops like wheat, rice, and sunflowers, cotton remains easily the region's biggest export. In Uzbekistan, the government remains the only purchaser, and meeting cotton-production quotas remains a national obsession. Even as the old collective farms are privatized, the quotas persist, forced labor is still used to pick the crop, and farmers and officials can lose their land and jobs for failing to meet them. Many American clothes companies have banned use of Uzbek cotton, because of the near-slave-labor conditions.[5]

Cotton still consumes most of the region's water. Today the countries around the Aral Sea—Uzbekistan, Kazakhstan, Turkmenistan, Tajikistan, and Kyrgyzstan—occupy five of the top eleven places in the world ranking of national per capita water use. Turkmenistan is on top. The Aral Sea basin is very far from being short of water. The problem is the simply staggering level of water use to grow cotton.[6]

I traveled through Uzbekistan, the heart of the old Soviet cotton empire. I drove from its capital, Tashkent, in the east, along the old Silk Road through Samarkand and Bukhara, and then north following the Amu Darya through desert toward its delta and a final, fateful destiny with the sands of the new desert. What struck me was not the single-minded determination to convert water into cotton but the sheer chaos that had resulted.

I drove first through the arid Hungry Steppe, west of Tashkent, where around 2.5 million acres are cultivated on one giant treeless cotton farm. All along the road, broken water channels leaked their water into the ground, and a salty white residue on the top of the newly harvested fields revealed the toxic consequences of massive overirrigation and poor drainage. Maintenance of the irrigation network had largely broken down since the Russians left. Waterlogging has brought the salts in the soils to the surface, where they form a crust after each harvest. Half the fields of Uzbekistan are salinized. The only practical way of removing the salt is to apply yet more water each spring to wash the salt into the drainage ditches that surround every farm.

So farmers are stuck on a treadmill, applying more water to grow their crops, which poisons the soils with ever more salt, which can be removed only with yet more water. In many areas more water is now used for flushing salt from soils before the cotton is planted each spring than for irrigation itself. But still the salt accumulates. Huge areas of land are being abandoned. Cotton yields are crashing on the fields that struggle on, so incomes are falling too. The closer I got to the delta the worse the crisis became.

Politically, the delta is within the autonomous republic of Karakalpakstan. Until the arrival of Commissar Cotton, the Karakalpaks were a nomadic cattle-herding and fishing race, whose name means the "black hat people." Many of them still wear black hats, though usually peaked caps made of cheap Russian leather. The Soviets and Uzbeks have both treated the Karakalpaks with disdain, using their "autonomous republic" for developing biological weapons and housing the notorious Jaslyq concentration camp.

In Karakalpakstan, I met officials sitting in drafty offices in their overcoats, their cold fingers leafing through huge ledgers as they attempted to manage the ramshackle irrigation systems and divide up the salty drainage water delivered to them by their upstream neighbors. Two-fifths of all the fields of Karakalpakstan—half a million acres—have fallen out of use since the 1980s, partly for want of water and partly because the soils are clogged with salt. Still they work to maximize production of the crop that is killing their land, emptying their sea and wrecking their climate.

Once, the Aral Sea moderated the harsh desert environment here—cooling summers, warming winters, and ensuring rainfall. Since it disappeared over the horizon, the summers in Karakalpakstan have become shorter and 5 degrees Fahrenheit hotter, the winters colder and longer. Every farmer I spoke to said that he used to plant his cotton in March, but now it was May or even June. "It's not warm enough to grow cotton anymore," one told me. Rainfall has declined, too, and the region is ravaged by dust storms whipped up on the exposed bed of the old sea. The dust carries with it an alarming cocktail of farm chemicals taken to the sea in past decades in drainage water. They include long-lasting pesticides like lindane, DDT, and phosalone.[7]

Alarming though the spread of pesticides on the wind may be, most researchers believe that there is a still more devastating threat in the dust storms—salt. The stuff is everywhere in Karakalpakstan. There is no escape. It comes on the wind, down the irrigation canals, and through the pipes carrying drinking water from reservoirs; it is in the vegetables the Karakalpaks grow in their gardens and the fish and birds they catch out on the delta; it is left behind on the soil surface by the irrigation process. Salt destroys the perilous productivity of the land, uses up precious water in flushing it out of soils, creates poverty, and ultimately kills the people themselves. Salt is the true tragedy of this land. Worse than the poverty, worse than the water shortages, worse than the pesticides, the land and its people are being poisoned by salt.

This is not a disaster that grabs you as you walk down the streets of Nukus or Muynak or the countless flyblown towns and villages across the delta. The people are too familiar, in their pullovers and shabby black jackets and flat caps. There are no refugee camps or soup kitchens. But probe a little and you find a blight that cuts right through a society that once prided itself on its order and ability to provide for all. And it affects everyone.

I talked for a long time to Oral Ataniyazova, a gynecologist and campaigner for the health of her people. Ataniyazova grew up on the delta, where her father was first secretary of the local Communist Party. Now she is famous for her work exposing the way that the Karakalpaks have been poisoned by Commissar Cotton. International acclaim has included the Goldman Environmental Prize awarded in 2000. She could have fled this blighted land but decided to return to work full-time for her people. Salt, she said, has turned Karakalpakstan into a nation of anemics. Among the eight hundred thousand women in the tiny republic, 97 percent suffer from anemia, five times the rate in the early 1980s, three times that elsewhere in Uzbekistan, and among the highest rates in the world. "All of our women are sick—and so are all our newborns," Ataniyazova said.

Anemia causes a disturbing number of pregnant women to die from hemorrhages during pregnancy and childbirth, and most of their babies are born anemic. At the maternity unit of the Kanlikol hospital, a nurse told me, "The children are slow and get sick a lot. We also have lots of birth defects, especially of the mouth, legs, and hands." In recent years, one in every twenty babies has been born with a defect. Infant mortality has been the highest in the former Soviet Union, and Karakalpakstan has the highest rate of cancer of the esophagus in the world, and unusually high rates of other cancers, immunological disorders, kidney and liver diseases, tuberculosis, allergies, and reproductive pathologies, said Ataniyazova. Average life expectancy is just fifty-one years.[8]

Later, I met a group of farmers meeting in Chumbai, one of the towns on the delta. These aging, weather-beaten men told me that nobody escapes the salt. "We get salty water, put it onto salty land—and then the winds bring another layer of dust as well," they said. They were certainly not experts at diagnosing medical conditions, but when I asked if they saw any health effects among themselves and their families from this toxic environment, they all nodded. My interpreter tried to shut off the conversation. It was too difficult for them to talk about, he said. But the manager from a state farm rose slowly to his feet, as if pronouncing at a funeral. He clutched his hat in front of him and looked me in the eye. "You can see it in the faces of everyone living here," he said. "We are all affected. There is a lack of good blood. All the women have it, and it is worse in pregnancy. Many children are born deformed here. In the drought years, lots of people died, especially the children. My daughter and son were both in the hospital for six months. Every family has someone like that." There was more nodding, but nobody would say any more. This was private grief among old men with sick wives, sicker children and grandchildren, failing crops, growing poverty, and a poisoned land.

So to the sea itself, or where the sea once was. I drove on to Muynak. Back in the 1950s, the sea still lapped the shores of Muynak. Soviet filmmakers portrayed the heroism of fishing fleets that trawled the Aral Sea from its harbor. The ships caught 50,000 tons of sturgeon, carp, and bream each year. Ferries sailed from Muynak to Aralsk, its companion port 200 miles away on the north shore in Kazakhstan. But there is more sand than sea between Muynak and Aralsk today. Muynak last saw the sea in 1968. The corroding remains of the last trawlers are stranded on the floor of the harbor. I was met by the manager of the former fish-canning factory. He showed me his empty warehouses and spoke of his dreams of reviving the plant if the sea ever came back.

Once, the townsfolk told me, the balmy climate meant they could go swimming in the sea in November. Now, with the loss of the moderating influence of the sea's waters, they have their overcoats on before the end of October. The population back in the 1960s was over forty thousand; today, after a mass flight of what might be called ecological refugees, it is around ten thousand. The sole income for many families is the pensions of their senior members. There was still a hotel. The rooms were dusty from lack of use, and when we asked for a meal, they suggested we go out and buy some food, which they would cook.

Muynak seemed like the end of the line. Yet beyond it, a few miles along what must once have been the coast road, was the small town of Uchsai, which means "tiger's tail"—a name it got because it sits at the very end of a peninsula that once jutted out into the Aral Sea. Back in 1995, on my first visit there, I found it one of the most depressing places I had ever been. On my return it was even worse, like a ghost town out of the Wild West. Its population had fallen from ten thousand to a mere one thousand, and few of these ghostly, emaciated people stayed for any reason other than lethargy and lack of an alternative. The town's one source of employment was a fish-smoking yard that limped on in 1995 but had closed down two years later. The piped water supply for the town was cut off, replaced by twice-weekly visits from a water tanker from Muynak.

I met two elderly schoolteachers in the town's only street. "The sea left here in 1961," one of them remembered. "It came back in 1966 and 1968. But that was the last time we saw it." Now they had only the sand and the salt. When I first came here, the workers at the fish-smoking plant told me that they expected the sea to come back one day. "It has dried up before and returned," one explained. Now, the women said, it was gone forever. A couple of boys played football listlessly in the road. "Most of our children are sick here," said one of the teachers. "They have anemia. You can tell. They are weak and slow and cannot study well." She blamed the sand and the salt in the air, the poverty and the lack of fresh food. "We used to grow vegetables in our gardens when we had piped water," she said. "But we can't do that now." There have been epidemics of TB and birth deformities in Uchsai.

You could be forgiven for thinking that the government in Tashkent had no interest in this forgotten corner. Not quite. A third of a mile down the road, I spotted a gas flare. They were drilling for gas beneath the Aral Sea bed and had built a gas compression station right outside Uchsai. In 2010, it was announced that some gas had been found.[9] But the large, fenced-off complex brought no benefits for the people of the town. "They never stop, they don't talk to us, they fill our town with gas smells and give us no compensation. They don't even buy things in our shop," the teachers complained. "But their trucks are destroying our road. When the snow falls in winter, the water tanker won't be able to get here. Then what will we do?"

On the way back from Uchsai, as we bounced and swerved among the potholes along the isthmus, we drove past the notorious ship's graveyard, where rusting hulks are blasted by sandstorms. For many, this scene has become a symbol of the disappearing Aral Sea. But what I remember is the betrayed people of Uchsai, lined up in the main street with old milk churns, awaiting the water tanker. When the river ran dry, the tide really went out for them.

Fountains are everywhere in Turkmenistan, the closed desert state of Central Asia. President Saparmurat Niyazov, who ruled here until his death in 2006, loved them. Fountains festoon the front of the president's palace in the capital, Ashkhabad, and line the road from the airport. Profligate use of water has long been the ultimate status symbol here. If you have it, you flaunt it. In that respect, if no other, Turkmenistan resembles Phoenix, Arizona.

In terms of rainfall, Turkmenistan may be one of the driest states on earth—a vast, flat stretch of desert extending north from the mountains of Iran deep into the wasteland of Central Asia. It has no rivers to speak of and few underground aquifers. Yet, preposterously, it has so much water that it has become far and away the highest per capita water user in the world. In a year, the country gets through 4 acre-feet of the stuff for each of its five million inhabitants. That's enough for each of them to flush a toilet twice a minute, night and day, all year.

Almost all the water that festoons Turkmenistan comes down the Karakum Canal, the Soviet-built artery through the Karakum Desert that empties the Amu Darya in Uzbekistan. Up to half a mile wide, the Karakum Canal is the largest irrigation canal in the world and the third-biggest waterway of any sort in Central Asia. It carries between 20 and 30 million acre-feet of water a year to water Turkmenistan's huge areas of cotton fields. The canal is now a decaying relic of Soviet times, increasingly silted up and losing half its water to seepage and evaporation. It still takes half the flow of the Amu Darya, just as it has for half a century—since Soviet leaders in Moscow decreed that Turkmenistan could have equal shares of the river with Uzbekistan, even though the Turkmen population is only a little over a quarter that of its bigger neighbor. Nothing, of course, was left for the Aral Sea.

Things are getting more bizarre. The country's erstwhile president, who styled himself Turkmenbashi, or "father of the nation," ran a paranoid, megalomaniac regime. Before his death, he announced his master plan to, as he modestly put it, "change the destiny of Turkmen people for generations to come." Niyazov wanted to create a vast new sea in the middle of the Turkmen desert. Not content, it seems, with his major role in destroying the Aral Sea in Uzbekistan and Kazakhstan, he would create a new sea on his territory.

Construction of the new lake, funded by billions of dollars in revenues from the country's natural gas fields, eventually began in 2009, under the tutelage of Turkmenbashi's successor, Gurbanguly Berdimuhamedow. It continues. The idea is to collect all the drainage water flowing off the country's immensely inefficient irrigated cotton fields and take it down a network of new canals to the Karashor depression in the northwest of the country.[10] Once full, the Golden Century Lake will cover around 750 square miles and provide water to irrigate another 1,600 square miles of fields—enough to grow an additional half a million tons of cotton.[11]

That remains the plan. There are a few problems. For instance, the long unlined drainage canals will lose water to leaks and evaporation on a vast scale. Up to half the water entering the canals is likely to disappear before it reaches the lake. If its surface area ever gets as big as promised, the lake will lose up to 6 million acre-feet of water a year to evaporation—almost a third of the water that Turkmenistan is currently entitled to take from the Amu Darya. This is a megalomaniac's mirage in the desert—a fitting encore to the destruction of the Aral Sea. There may be a little water left to grow cotton. But probably not much.

IX

When the rivers run dry . . .

we try to catch the rain

29

HARVESTING THE MONSOONS

LIAN JIANMIN—a smiling, worldly, chain-smoking raconteur—was telling me his life story. "I was a seventeen-year-old just out of high school in Tianjin in the east. During the Cultural Revolution in the late 1960s, when students were sent to the countryside to learn about peasant life, I went to work in a village in the hills of Gansu, in western China. I saw they all had these cellars under their houses, where they collected the rain from their roofs." Many of the cellars were very old and had silted up, he said. People preferred water from wells. Then there was a drought, and the wells went dry. Villagers with cellars began using them again. Those without had to walk 5 miles or more to find water. Afterward, everyone started digging out their cellars to catch the rains again.

The sandy hills of Gansu are one of China's poorest and most desolate regions. The wells are prone to run dry and their water is often salty. But they do have a little rain and there is a long tradition of catching that rain before it runs away. Some of the water cellars are said to be eight hundred years old. Lian said he met a farmer who was eighty years old: "When he rebuilt his house, which he'd lived in all his life, he discovered a cellar underneath that still contained water. And the quality of the water was very good." Lian became so captivated by the structures that he decided to make their revival his life's work. Now an engineer with the province's water resources bureau, he has entertained a growing galaxy of foreign water experts who come to see how it is done. "We had people from thirty countries here last year," he told me.

An estimated two million people in Gansu today get most of their water for drinking and growing crops by harvesting the rain. The cellars catching water for domestic use have been augmented by an increasing number of cisterns dug in the fields to provide water for growing crops. Big enough for a man to stand in, they usually have a distinctive bell shape and typically hold around 15,000 gallons of water. They are usually lined with cement, though some say the old cellars lined with mud bricks give better-tasting water.[1]

Every household that I visited had at least one cistern. Sheep farmer Zhujia Zhuang had piped water in his home, but still kept a backyard cistern to water a small vineyard and a second one hooked up to a solar heater in a shower cubicle. An Yueying, also a farmer, showed me around her tiny but immaculate farmhouse. She relied on the rain for all her water, she said. A cistern outside caught water from the gently sloping yard, and another collected water running off a hillside and down a small concrete channel. "I built them myself about twenty years ago," she said. She was growing alfalfa for her cattle in a field, and there were pear trees and vegetables in the yard. It was the end of the dry season, but there was still water in both cisterns.

Was the water safe to drink? Usually it was all right, she said; but to be sure, she boiled it first. As we spoke, she filled a battered old black kettle from the cistern in the yard and hung it up above a giant shiny dish pointed toward the sun. I had at first taken this for a big TV satellite dish, but in fact it was concentrating the sun's rays to heat the kettle. A while later, the kettle came to the boil.

Later, over lunch, I met Zhang Zhenke, the director of the Dingxi County water conservancy bureau. Most of rural Gansu drank from cellars, he said. "We have two hundred thousand water cellars in this county alone. A quarter of them are privately built, and the rest were initiated by the government." Some people, not content with taking water from their roofs, laid plastic sheeting on a small corner of their fields as a catchment area, he said.

In many parts of the world, water officials would be embarrassed to admit that only a tenth of their customers were connected to piped water, but Zhang was proud of the achievement. "We already have more rainwater harvesting here than anywhere else in China, probably the world. Without the cisterns, people could not live here," he said. Cisterns were not a historical relic; they were the key to the future prosperity of the region. The policy now was for every household to have a concrete catchment of 1,000 square feet to fill two cisterns. "Eventually we want four for every household," he said. "There would be one for drinking water for a normal year, one for a drought year, and two for irrigating crops."

In China, it was Chairman Mao and his Cultural Revolution that began the revival of the ancient tradition of rainwater harvesting. In India, it has been a mixture of swamis and scientists, schoolteachers and even policemen. Haradevsinh Jadeja is a retired Indian police officer. He loves playing cricket for his village team. When we met, he had a broken arm caused by a fast bowler from the next village. But win or lose at cricket, his village of Rajsamadhiya, in the backwoods of Gujarat in western India, always excelled at water.

His fellow villagers told me Jadeja was a diviner of water, and I could see what they meant. Around the village, he had transformed a desert-like landscape of desiccated fields and empty wells into a verdant scene of trees, ponds, full wells, and abundant crops. With no piped water, most of the other villages in the area relied on tankers to provide drinking water for much of the year. They had little left to irrigate their crops outside the short monsoon season. But in Rajsamadhiya, "we haven't had a water tanker come to the village for more than ten years," said Jadeja. "We don't need them."

He worked this magic by catching the monsoon rains before they drained away. Not, like the Chinese, in purpose-built cellars and cisterns, but in ponds. The villagers didn't use the water directly from the ponds. Instead, they allowed it to percolate into the soil to refill underground water reserves and so replenish their wells. "There is no more rain than before. We just use it better. We don't let it wash away," Jadeja said. The village had twice as much water as before. Wells found water at only 20 feet down, whereas in neighboring villages the water had to be hauled up from more than 100 feet.

The heart of the village appeared conventional enough: a gaggle of single-story houses leading from a small square out toward the fields. On the paths, however, there were thousands of fruit trees, where most other local villages were treeless. Under their shade were piles of mangoes and watermelons. Out among the small fields growing wheat and vegetables and groundnuts, there were the ponds—lots of them. "We have forty-five water-collecting structures altogether," said Jadeja as we walked past a line of women washing clothes in a pond.

The ponds had been dug to a plan. They were arranged along the routes that the monsoon rainwater took as it drained through the village. Jadeja had redesigned the village's drainage to slow the water's passage long enough for it to collect in the ponds. The water then passed from one pond to the next in a slow cascade, seeping through the soil to refill the aquifer all the way. Jadeja's second innovation was to manage how people use the water. As a former policeman he had some authority, and though all decisions were taken by the village council, his word had obvious force. In fact, he told me quietly, the police and other authorities never come to the village now. They leave it all up to him. Under Jadeja's law, nobody was allowed to take water directly from the ponds, and farmers have been banned from growing the thirstiest crops, like sugarcane. "There is no point in catching more water if we only waste it," he said.

News about this remarkable village has spread around India and beyond. One foreign scientist brought satellite images of the village that showed hidden cracks in the geology through which water was flowing. Jadeja saw a chance to catch more water by plugging the cracks with concrete. But mostly, as in their pilgrimages to Gansu, water scientists have been citing Jadeja's water fiefdom to learn about rainwater harvesting, though he insisted that there was little to learn. "I am an uneducated person," he said. "I saw that people were leaving the village and I wanted them to stay. That meant finding more water. So I tried to catch the rain."

Jadeja has tapped into an old tradition and developed it. In India, you can still see abandoned ponds and lakes dotted across the countryside and on wasteland in cities. Until the early nineteenth century, much of the country was irrigated from shallow mud-walled reservoirs in valley bottoms, which captured the monsoon rains each summer. The Indians called them *tanka*, a word the English adopted into their own language as "tanks."[2] Most of the tanks were quite small, covering a couple of acres at most, and irrigated perhaps 50 acres. Farmers scooped the water from the tanks, diverted it down channels onto fields, or left it to sink into the soil to refill their wells. The tanks served other functions too. Some were stocked with fish. All were prized for the silt brought into them by the rainwater. Farmers guarded the slimy, nutrient-rich mud almost as much as the water. They dug it out to put onto their land and turned silted-up former tanks into new farmland.

Overall, across India, researchers estimate there are around 140,000 tanks, either still in use or abandoned. Tamil Nadu in the south has the most, approaching 40,000, covering several percent of the land surface of the state and still irrigating around 2.5 million acres. The system thrived until the British took charge in India. Though full of admiration for some of the grand Indian water structures on rivers, British water engineers largely ignored the village tanks, apparently not realizing that they were the way India fed itself. Tanks passed into a kind of forgotten nether land, used while they served a purpose but unrecognized by officialdom and rarely repaired or cleaned out. As the British, and later the Indian government, promoted more modern water-gathering technologies, the tanks gradually fell into disuse. But today, as formal irrigation systems established on the Western model fail across the country, and as farmers pump from ever greater depths to retrieve underground water, the old tanks and other rainwater harvesting structures are starting to be restored.[3]

Every region has its own design. In the Thar Desert of northwestern India, tanks are called *johads* or *khadins*. A variant is the check dam, constructed across a small stream or in a gully to hold up the drainage of monsoon rain long enough for it to percolate underground. Often, charismatic individuals have been behind their revival. In Rajasthan, an Ayurvedic doctor, Rajendra Singh, learned traditional water-harvesting skills from villagers and began traveling the Thar Desert, where villagers were dying for want of water. The man they now call the "waterman of India" built his first *johad* in Gopalpur village in the mid-1980s. It took him four years to complete. "Luckily in 1988 we had very good rainfall. All the underground wells got filled because of the *johad*. The villagers were overjoyed," he said later.[4] Word spread. Now he has forty-five hundred *johads* collecting water in several hundred villages, forty-five permanent staff, and a grant from the Ford Foundation.[5] Mainstream water engineers assembling at their annual gathering in Sweden in 2015 awarded Singh the Stockholm Water Prize.[6]

Arid India seems to be full of "watermen." The river Arvari in Rajasthan had been dry for decades. The fields along its banks were parched, the people were leaving. Then a local youth, Kanhaiya Lal Gurjar, began recruiting village women to construct simple earthen check dams in the beds of seasonal streams that once fed the river. As the water tables rose, local wells began to return to life and, as if by magic, the river itself eventually returned. "The villagers have better knowledge about water conservation in their areas than most engineers," says the author of a study of *johads*, Meera Subramanian.[7]

Indian hydrologist Tushaar Shah, who for many years ran an international groundwater-research center in Gujarat, is a fan of rainwater harvesting. He says the success of these efforts owes as much to sociology as hydrology. Few individual farmers can successfully catch their own rain and store it underground; it would quickly dissipate into the wider aquifer. But when an entire village does it, the effects are often spectacular. Water tables rise, dried-up streams flow again, and with more water for irrigation, the productivity of fields is transformed. He sees the rainwater-harvesting movement as "mobilizing social energy on a scale and with an intensity that may be one of the most effective responses to an environmental challenge anywhere in the world." It is, he points out, completely autonomous from government. "It emerged on its own, found its own source of energy and dynamism, and devised its own expansion plans."

~~~

There has never been a blueprint for catching the rain. You can catch it on rooftops or on agricultural terraces, by diverting flash floods into ponds, by blocking gullies, or by putting low earth embankments across hillsides or around individual plants. Harvesting the rain was once a worldwide technology and every locality had its own system. English country houses caught the rain from their roofs to water the flowerbeds. So did palaces from India to the Middle East. African villagers harvested water from the leaves of trees in their kitchen gardens. Native Americans erected low stone walls and weirs made of brushwood across the Sonoran Desert to divert storm waters onto their fields. For small tribes and isolated communities in particular, rainwater harvesting has always made much more sense than the larger river-diversion structures of "hydraulic civilizations."[8]

Many countries in the Middle East have continued harvesting the rain in defiance of Western engineering norms. Jordan constructs earth dams in the desert to increase soil moisture for groves of olives, pistachios, and almonds. Yemenis captured floodwaters in wadis for centuries and exported the idea in the early twentieth century to the East African states of Eritrea, Ethiopia, and Somalia, where this is now the dominant form of irrigation. Tunisia still grows a hundred thousand olive trees with *meskats*, rain catchments as big as tennis courts and surrounded by earth embankments. A similar system continues to water orchards in parts of Afghanistan. In the desert sands of Egypt, the ancient Bedouin technique of cutting cisterns into the rocks to catch and store water for drinking and irrigating vegetables has undergone a dramatic revival. Thousands of new cisterns have been cut.[9]

The inhabitants of small islands surrounded by salty seas know better than most the perils of a finite supply of freshwater. Madeira, a holiday destination in the warm waters of the Atlantic Ocean west of Morocco, has few rivers. But it does have a network of 1,500 miles of beautiful stone water channels, known as *levadas*, that were constructed by African slaves for Portuguese sugar plantations. The British territory of Gibraltar, on the southern tip of Spain, which is politically if not quite geographically an island, has one of the most intensive rain-collection systems in the world. It channels water from collecting areas on the giant rock that dominates the enclave into tanks excavated into the rock. Across the Atlantic in Bermuda, new houses are required by law to include equipment for collecting rainwater from their roofs.[10]

If any moment was pivotal to the rediscovery of rainwater harvesting in the West, it was when an Israeli archaeologist, the late Michael Evenari, of Hebrew University of Jerusalem, stumbled on hillside water channels in the Negev

Desert. In the 1970s, he was excavating settlements of the Nabateans, who con-trolled the caravan routes from Arabia to Rome via Mediterranean ports in Gaza around two thousand years ago, trafficking in gold, ivory, spices, frank-incense, and myrrh. This was big business. The Nabatean kings were probably the three kings of the Nativity story, and they ran six imposing desert cities, including the famous "pink city" of Petra in modern Jordan. How, Evenari wondered, were they fed and watered?[11]

As he excavated, Evenari realized that these traders had caught the rain as it fell in occasional storms on the hills. He found extensive evidence of low walls on hillsides that captured and channeled the rains. Hydrologists estimate that a single acre of Negev Desert can capture as much as 100,000 gallons of water over a year. Cleverly managed, this would have been enough to grow olives, wheat, and vines. He was so struck by the idea that he re-created an ancient Nabatean farm below the ramparts of Avdat, one of the ruined Nabatean cit-ies in the Negev. When I visited his Avdat farm, it hadn't rained for six weeks. All around was barren wasteland. But on the farm the soil was damp, wheat was growing, and almond and pistachio trees were in leaf. "We have tried to grow the crops mentioned in the Bible," said Pedro Berliner, a researcher who took over Evenari's work. "And most of them will grow here in the desert if we harvest the rain."

It was a stunning sight. As I have spoken to practitioners of rainwater har-vesting around the world since, I have discovered how important the Negev experience has been in their work. Several Indian advocates made pilgrimages to Israel back in the 1980s. It was around then, too, that Bill Hereford, of the development agency Oxfam, came across Evenari's desert farm while on a sab-batical from his work in Burkina Faso. On his return, he encouraged African villagers to try to revive abandoned desert fields on the edge of the Sahara by laying low stone walls on the hillsides. It worked. The walls captured water long enough for it to soak in and nourish crops.

The news spread across West Africa. Thousands of acres across the arid Sahel have crops and trees growing where before nothing grew. There has been a 70 percent increase in yields of local grains such as sorghum and millet, as well as more trees for firewood and more grassland for livestock, says Chris Reij of the World Resources Institute in Washington, DC. Rainwater harvesting is the leading cause of a widespread greening of the Saharan margins in the past twenty years, he told me.[12]

There have been other notable instances of the cross-fertilization of the rainwater-harvesting idea. In the Machakos district of Kenya, the Akamba tribe was in a bad way sixty years ago. A British administrator called the hills where they lived an "appalling example" of environmental degradation: "The inhabitants are rapidly drifting to a state of hopeless and miserable poverty and their land to a parched desert of rocks, stones, and sand." Rather than accept their fate, the Akamba people sought out new ideas. Elders who had served in the British colonial forces in India during the Second World War remembered seeing *tanka* and hillside terraces. They decided to try the idea back home. They dug ponds and terraced hillsides to such effect that, despite a huge surge in the Akamba population, their farms are producing more, their people are wealthier, and their landscape is greener. Researchers call this transformation "the Machakos miracle." Desertification has been put into reverse through catching the rain.[13] Most recently, South African researchers have begun promoting the technique for small farmers there.[14]

What is the potential? "Millions of acres of the dry parts of the world were once used for water harvesting. In the twentieth century there has been a steady decline, but the twenty-first century is likely to see a huge revival," believes Dieter Prinz, formerly of the University of Karlsruhe in Germany. Social and economic conditions may have changed, "but the advantages of water harvesting remain valid, and farmers in dry areas will have to use them if they want to be able to master their future."[15]

30

OMAN

Unfailing Springs

IT WAS 47 DEGREES CELSIUS. Make that 117 degrees Fahrenheit. In mid-May 2015, when I was there, the desert of northern Oman was the hottest place on the planet. But in the shade of the oasis, the temperature was dramatically cooler. In his white robes, my bearded host Ali Al-Muharbi beamed as he showed me around the date palms, all irrigated by water gurgling down a channel dug many centuries ago to exploit underground water in the nearby Hajar Mountains.

Al-Muharbi was the water manager of the village of Al-Farfarah, about half an hour's drive into the mountains from the Omani capital, Muscat. As we toured his cool oasis, he described how a tunnel more than a mile long delivered 200 gallons of mountain water every minute into the network of channels that irrigated the hundred acres of date palms around us. We watched as field workers blocked and unblocked the small channels, each only a few inches wide, to divert water between the family plots. Bananas, fruit trees, and winter leafy vegetables all grew under the shade of the palms. "I don't know when this started," he said. "But it goes back for centuries; it is as old as the village. There would be nothing here without it."

Oman is a desert country on the shores of the Arabian Sea. These magical waters conjured from the most arid land imaginable are called "unfailing springs." Even in the worst droughts, their flows persist. The tunnels are pre-Islamic feats of hydraulic engineering that remain the only water supply for many villages. Even large towns owe their siting to their perpetually flowing waters. The systems for running them remain independent of the state and depend entirely on ancient community committees. The tunnels and their waters are known individually as *falaj* and collectively by the plural *aflaj*. They make the deserts bloom.[1]

"The *aflaj* may be the most ancient community-run systems for managing water in the world," according to Slim Zekri, a water economist I met at the Sultan Qaboos University in Muscat. Some connect to natural springs; some capture water in gravel beneath the beds of wadis; but the largest and most "unfailing" are those that are connected to tunnels that bring water from deep in the mountains. "Some of the tunnels are big enough for you to walk in," he said, "and long enough that they need shafts from the surface every few hundred yards so people can get down to clean them."[2]

They are not unique to Oman. Similar tunnels were dug long ago elsewhere across the Middle East. The origins of the technique lie in Iran, but they turn up along the southern shores of the Mediterranean as far as Spain, and east through Iraq into Afghanistan and western China. They have many names, including *qanats* in Iran and *fogarra* in North Africa. Only in Oman have they remained to this day the dominant source of water across large areas.[3]

Al-Muharbi's *falaj* in Al-Farfarah village is typical. The system has fifty owners—descendants of the people who first put up the money or dug the *falaj*. For six days a week, the owners each have established rights to water coming down the channel. Timeshares owned by families range from half an hour to fifteen hours, which is enough to irrigate five hundred palm trees. The shares are bought and sold, as some families subdivide land and water for the next generation, while others buy out absent owners. In Al-Farfarah, a permanent right to thirty minutes of *falaj* flow a week costs a bit over $3,000.

The water on the seventh day is open for purchase by any of the ten village families without an established share, or by share owners who want more water. Al-Muharbi used to oversee regular auctions of this "spot market" water in front of the village mosque. Prices varied according to the time of day and the season, but averaged about $5 for thirty minutes of water—a fraction of the cost of the desalinated seawater now being brought by pipe to some villages for drinking. Nowadays auctions are rare, he said, but many villagers still purchase this "seventh-day" water. The fees pay for Al-Muharbi's work, running repairs, and for charitable activities in the village, such as buying equipment for the village school.

Zekri, the water economist, thinks these simple water markets are the key to the long-term success of the *aflaj*. "The existence of private water rights that can be traded makes the system more efficient and allows the community to be self-reliant," he told me. "They could offer important lessons for the world in how to manage scarce water reserves."

On the evidence of my visit to Al-Farfarah, the complex water-sharing systems appear to be widely adhered to. The water from the mountain tunnel first enters the village, where it is accessed for drinking and domestic use; then it goes to the mosque and school and to communal bathing cubicles and a washing area, though soap is banned. Water from these sources all returns to the main channel and heads for the fields, where it is shared out among its owners.

I watched as farmworkers meticulously kept to their schedules, blocking and unblocking the narrow irrigation channels using old rags held in place by stones. One water owner, standing astride the channel irrigating her plot, berated her workers for not getting water to one palm tree. It could have been a scene from a thousand years ago.

But things are changing in the villages.

One example is how they time the water distribution. Al-Muharbi, who was sixty-five years old when we met, was keen to show me the traditional method used when he was young. Standing in the middle of the village square, he held a tall stick aloft. As its shadow moved and reached marks gouged on the ground, his predecessors would have sent out orders to the fields for which channels should receive water. At night, he remembered using the rising and setting of the stars as a clock.

I later discovered that British researcher Harriet Nash of Exeter University toured the villages here as recently as 2005, documenting for a PhD thesis how eight villages still distributed water according to the stars. It was by then unique to Oman.[4] The practice has since disappeared. Nowadays, day and night, the villagers need nothing more than their fancy watches or mobile phones.

There are other changes. With desalinated seawater now supplied by the government to many villages for domestic use, day-to-day survival no longer depends on *falaj* waters. Some locals no longer seem to know the rules about using the water. Traveling the villages, I saw signs in Arabic and English warning against bathing in water channels and banning the washing of cars with *falaj* water.

With young Omanis moving to the cities for work, the old men left in charge often employ contract workers from South Asia to work their farms and maintain the water systems. As we walked the channels of his *falaj*, Al-Muharbi was silently followed by his Bangladeshi factotum, Mohammed Islam, who carried a small spade to remove silt from channels. Islam told me he had been living in the village doing this work for five years now. He was enthusiastic. But there was an unmistakable sense that old indigenous traditions and skills are dying.

Until the 1970s, members of a tribe called the Awamir, who are based near the town of Izki, not far from Al-Farfarah, traveled the country repairing collapsed tunnels. They gave up because there was no money in it. Now foreign laborers do the repairs. They are brave enough to crawl down dangerous collapsing tunnels. But they often lack the traditional skills, said Abdullah Al-Ghafri, who has been researching *aflaj* for twenty years, when we met in his office at the University of Nizwa. Without due care and attention, the unimaginable is happening. Even as their fame spreads, the *aflaj* are failing.

UNESCO designated five of the best *aflaj* in Oman as World Heritage Sites. They have plaques and receive royal visits.[5] "Prince Charles came here from England," one proud manager told me. When I visited one of the famous five, Al-Malki near Izki, I discovered that only two of its seventeen channels, which once extended for a total of 9 miles, were still in use. Their water was little more than a dribble.

Back in his house in Al-Farfarah, I sat cross-legged on the floor with Al-Muharbi, eating the year's first crop of dates and oranges, washed down with tiny cups of tea. I asked him what he thought the future held for his *falaj*. His smile for once slipped. "The younger generation aren't interested," he said with a hollow laugh. "People still pay for the water, but the shares rarely change anymore." Worse, "the flows out of the mountain are declining even here. And it won't be maintained. It could dry up one day. Perhaps soon." It was a fatalism that could yet consign the unfailing springs to the history books.

Some blame the decline in flow on climate change. But, in truth, the droughts here are no worse than they have ever been. The real problem in Izki, Al-Farfarah, and elsewhere, is pumping. Farmers are sinking boreholes in upland areas where the *aflaj* channel the underground water. Their pumps are lowering the water tables until they are below the level of the *aflaj*, leaving the tunnels high and dry. *Aflaj*, which flow entirely by gravity, are self-limiting: they take water from the aquifers but cannot empty them. Pumped boreholes have no such limits—they keep sucking until all the water is gone. They are sabotaging the *aflaj*.

The pity is that, when their water supplies beneath the mountains are not disrupted, the *aflaj* still work remarkably well. Seen from the air, the green splashes of date-palm plantations clearly mark where they still water the land. Most of those that I saw were well maintained. Their managers regularly organized work forces to keep the channels unclogged. Hydraulic engineering features such as siphons, storage tanks, and raised aqueducts were in good working order. The government in Muscat is keen to try and protect them.

An inventory in the 1990s found just over four thousand *aflaj*, of which three thousand then remained in use. They delivered about 600,000 acre-feet and irrigated some 64,000 acres, up to half of all the country's fields. In an effort to protect them, the government imposed a ban on sinking new boreholes within 2 miles of underground *aflaj* water-collecting zones. But existing boreholes were allowed to remain. So, by 2014, another quarter of the *aflaj* had stopped flowing and a third were in sharp decline.

Ministers have promised a program of restoration, bringing in engineering companies to clean out and line the water tunnels.[6] Al-Ghafri fears that lining the tunnels could make matters worse, however. The problem is that the tunnels are not simply conveyors of water. They are also receiving chambers for water percolating from the rocks above. The Awamir knew this and rarely lined their tunnels. But unskilled repairers often line the tunnels with concrete in places where the tunnels once received water during wet times. The lining seals the tunnel off, killing what they came to cure.

Al-Ghafri wants to set up a research center to increase knowledge of the hydrology of *aflaj* and preserve the Awamir secrets of traditional management. "We need to interview the last Awamir tunnel diggers before they are gone," he told me. He also wants to investigate the conservation value of the *aflaj*. These systems are the only permanent sources of water in many areas of Oman, yet their ecology is largely unknown. They could be harboring Omani equivalents of the blind white fish that famously inhabit Iranian *qanats*.[7]

Better hydraulic engineering may not be enough. The key to keeping the *aflaj* and the fields that they irrigate may be economic, says Zekri. Much of *falaj* water is wasted because it flows constantly to the fields, whether or not it is needed. Farmers routinely overirrigate. The answer, he says, is to install gates where the tunnels exit the mountains so the flow can be stopped when it is not needed and kept for another day. He is also looking for a village that will try replacing traditional timeshares with smart meters to encourage water conservation.

Meanwhile, we may be witnessing a classic tragedy of the hydrological commons. Private greed, in the form of boreholes and pumps, is wrecking a sustainable and collectively managed water system that archaeologists say has been delivering water in this area for five thousand years. The oases are drying up. The date palms are crashing to the ground. Waterless villages may soon be abandoned.

31

IRAN

From Under the Mountain

AFLAJ, OR *QANATS,* are one of the great engineering treasures of the Middle East and the arid regions that its civilizations have conquered. In much of the region, rain falls only sporadically and mostly in the mountains, where there is little flat land or soil for farming. The water swiftly percolates underground, accumulating in the stony slopes that flank the eroding mountainsides. If the water surfaces naturally at all, it does so in small fluctuating springs on valley floors. In such circumstances, it was the Persians, in modern Iran, who first learned to excavate these small, tantalizing springs, chasing the water back into the hillside by digging tunnels. They found that the farther they dug, the more water the springs delivered and the more reliable the flow was. What probably began as a smart idea to get through a dry spell turned into the wellspring for farming societies across a huge swath of the Old World.

Most textbooks on water technologies make little mention of *qanats,* yet they are engineering marvels, dug straight and true, at a fixed gradient for their entire length, whatever the terrain above. The scale of their construction is truly extraordinary. In Iran, there are an estimated fifty thousand *qanats,* mostly dug more than three thousand years ago during the heyday of the old Persian empires. The tunnels often extend for many miles into the hillsides. Assembled end to end, they would reach two-thirds of the way to the moon. If these structures were above ground, they would surely be regarded as among the great wonders of the world.[1]

Once, *qanats* were the main source of water in Iran. Until the 1930s, dozens of them supplied the capital, Tehran. One gushed into the grounds of the British Embassy. Tabriz, Shiraz, and many other cities ran on *qanats.* One, supplying Isfahan, was more than 45 miles long. Some had such strong flows that the Persians built water mills on them to grind grain. As recently as the 1960s, their total discharge was around 485 million acre-feet a year, equivalent to the flow of eight Nile Rivers. But, even more than in Oman, modern water pumps have lowered water tables, drying up the tunnels. The most recent estimate of productive use, made in 1998, was around 8 million acre-feet.[2]

In many places they are now little regarded, and the skills that made them are dying fast. But in parts of Iran they still form the backbone of the water supply. The city of Irbil and its hinterland remains a hotspot. The *qanats* here were dug by slaves of King Sennacherib, twenty-seven hundred years ago.[3] When an earthquake struck Bam in eastern Iran in late 2003, one of the first discoveries made by aid workers arriving in this remote town was that it was entirely dependent on *qanats* for water—and most of the tunnels had collapsed in the quake.[4]

In the hot desert of central Iran, the ancient city of Yazd relies on *qanats* up to 50 miles long to bring water from the snow-covered Mount Sir. They channel the water beneath the city streets like a subway system, supplying water to underground public water-houses, called *ab-anbars*. Yazd is a city of tiled mosques, carpet shops, and ancient Persian wind towers, which ventilate houses through turrets oriented toward the prevailing valley winds. Many of the water houses have wind towers attached. They are oases of cool air, cool water, and well-being.

Wherever the Persians went, they took the secrets of *qanats*. The technology spread along the Silk Road to Afghanistan and China, where *qanats* are called *karez*; through Arabia, where they are *aflaj*; and along the north coast of Africa, where locals call them *foggara*. *Qanats* watered Moorish Madrid and are still the mainstay of peasant agriculture in the Algerian Sahara. The 50,000-acre Ouled Said Oasis grows dates watered from *foggara* dug into the nearby hills. Recent excavations have found hundreds of *qanats* beneath the Libyan Desert, relics of the ancient civilization of Garamantes. They lie abandoned not far from the green pipes of Gaddafi's Great Man-Made River.[5] In Kurdish northern Iraq, the great city of Sulaimaniya made heavy use of them. UNESCO found that many across northern Iraq dried up during a drought in 2005, leaving some one hundred thousand people forced to flee for want of water.[6]

Many more succumbed during the drought in northern Syria that began in 2007, a region that I had visited three years before. Shallalah Saghirah was a small, shabby village in northern Syria. Its name meant "little waterfall." Its inhabitants told me the story of their community and its water. The village had been founded by Musa Oqlah, a migrant from the south, who arrived in 1928. The land was deserted and dry, a frontier of dry Bedouin goat pastures. When he planted barley, the crop died. He was on the verge of moving on when he noticed a few sprigs growing around a seep of water. He started digging and found what seemed to be a well. He stayed and drew water from the well to irrigate his barley.

A decade later, Oqlah's well ran dry. He did some more digging in the gulley behind the village. What emerged staggered him. He found a chain of wells connected at their bases by a tunnel 1,300 feet long, which tapped underground springs up in the hills. Oqlah had discovered an ancient *qanat*. He and his sons cleaned out the wells and the tunnel and constructed a stone surface channel to take the water to the middle of the village. The *qanat* had flowed ever since, the villagers told me.

Shallalah Saghirah had about twenty households when I visited. All seven elders were grandsons of Oqlah. They had no electricity, and their only source of water was the underground tunnel. After a renovation in 2000, the village had doubled its discharge. It poured into a small reservoir at a rate of about a quart a second, as regular as clockwork, whatever the season. Ismail Hasoun, one of the elders, clambered like a mountain goat up the gulley behind the village to show me the seven wells along the tunnel's route. He said that near its head, the tunnel branched into several sections, apparently to access more of the aquifer.[7]

Hasoun's grandfather had divided up the water between his five sons. Each got a day of its flow in turn. The fields were still owned by their descendants, and the share-out continued much the same. They grew figs, mulberries, vegetables, and barley. "The other villages in the valley have to buy water from tankers or the government wells," Hasoun said. "Their water isn't as good as the water we get from the tunnel."

Researchers have mapped around 250 *qanats* across Syria.[8] One tunnel, supplying Turkish baths in the ancient Byzantine city of Aleppo, is more than 7 miles long. But Shallalah Saghirah may have been the only community where a *qanat* was the sole water source.[9] There were no other *qanats* in the surrounding villages in the Khanasser Valley. Or that's what people told me. But the local Bedouin had excavated underground cisterns up in the hills to provide water for their goats. Some still held water. The older men knew about them. One of them, Mohammed al-Issa, took me to see some cisterns up a long gulley near the town of Habs. It was a stumbling climb over rocky terrain, past abandoned shacks built by shepherds who once brought their animals there. Every few hundred yards there was a hole in the ground leading to a water-filled cistern, each connected by a tunnel. As we returned to the road, he pointed out a beautifully constructed stone channel that followed the slope down to a nearby village. It was dry, but it must once have delivered water from the cisterns. So Shallalah Saghirah was not alone after all. But as we followed the channel, we noticed men dismantling it and loading pieces of stone into a truck. They said they were taking it for building. Before our eyes, they were demolishing the *qanat*.

The story is poignant, especially given what has happened to the region since. As the drought worsened after 2007, there was a mass exodus of people from the area, because they had no water. They headed for Syrian cities, where unrest led to the growth of opposition to the Assad regime and ultimately the rise of Islamic State. What has happened to the *qanats* in Shallalah Saghirah and outside Habs since then, I am unable to say. But had the hundreds of old *qanats* still been in working order, it is at least possible that their water would have continued to flow through the dry years. If it had, then the exodus would have been prevented, and perhaps the civil war would never have happened.

One spring morning on the eastern Mediterranean island of Cyprus, I interviewed a man who claimed to be the last *qanat* digger there. To prove the point, sixty-five-year-old Yannis Mitsis took me to the orchards near his village, where he assembled a large wooden tripod, attached a pulley, and lowered himself down a hidden well into a tunnel that still fed water from the island's snow-clad mountains. The sun had already dried up the rivers in the lowlands, but his tunnel was running with water.

Back at the surface, Mitsis described to me how, as a young man, he had helped excavate new tunnels, which are locally called *laoumia*. The work was an ancient family occupation, though a dangerous one. His father had died years before when a rope broke. In the old days, Mitsis said, "the tunnels were the source of water here, and water was power." Nobody had dug a new tunnel since 1954, but he was kept busy doing repairs. Many orchards still relied on them, and on him. There was nobody to take his place. "My son is a criminal lawyer in London," he told me. "When I stop work, there will be nobody left."

There is almost nothing written about *laoumia*. I learned about them from a student who wrote to me after reading an article I wrote on *qanats*. There is a subculture surrounding such tunnels. From an article by environmental journalist Peter Bunyard, I learned that there are tunnels very similar to *qanats* but shorter and excavated from hard rock, in the hills of the West Bank. They are known locally as spring tunnels, and most villages on the West Bank have one somewhere. Most are used only as a last resort by poor Palestinians such as d Qot, whose story we heard in chapter 21. Many have been mapped by aeli geographer from Tel Aviv University, Zvi Ron.[10]

Almost everywhere, these tunnels seem to be dying. Yet they have hydrolog-qualities that should be in great demand. They are self-regulating, because like electric pumps, they can take water from the aquifer only to the limit of atural replenishment. As a result, if properly designed and maintained, they re, as they call them in Oman, "unfailing springs." They keep going in all but the worst droughts. They are a source not of conflict, like many pumped wells, but of social cooperation. The tunnels often emerge from underground at the home of the most important family in the village, but thereafter everyone is entitled to water. Typically, they have many owners, and the water is divided into hundreds or even thousands of timeshares. In Ardistan in central Iran, the division of waters at one *qanat* goes back eight hundred years to when Hulaga Khan, the grandson of Genghis Khan, ordered the share-out.

Can *qanats* make a comeback? In Oman, the government has spent money on rehabilitation and attempting to integrate them into modern agriculture. In the huge Turfan basin in the western Chinese province of Xinjiang, more than a thousand *qanats* dug during the Han and Qing Dynasties were renovated in the 1990s. They take water up to 12 miles from beneath the snow-covered Bogdashan Mountains, and reportedly still provide a third of the local water.[11]

Aid agencies are becoming interested. Often, abandoned *qanats* are the most viable new sources of water to supply refugees in conflicts. Mustang Valley, in the arid Pakistani province of Baluchistan, was once watered by 260 karez. After the valley was connected to the electricity grid and locals began to drill boreholes for water, the karez were abandoned. But during a drought in 2001, as Afghan refugees flooded across the border in the months after 9/11, teams of water engineers found they could still tap the old karez to supply the refugee camps.[12]

When American soldiers went to Afghanistan in 2001, they became suspicious of the many tunnels in the hills. Many believe that Osama bin Laden escaped the American dragnet in Tora Bora in the final days of the war there by fleeing down karez, many of which are wide enough to accommodate small companies of men. American engineers were amazed to discover that karez were indispensable to Afghan farming, watering as much as a sixth of the country's irrigated fields.

If anything, the importance of Afghan karez has grown since. In 2001 the aid organization Islamic Relief encouraged locals to renovate seventy-five karez in Helmand Province, where the Helmand River had run dry. Western aid agencies briefly decided to continue the work. The UN proposed constructing small check dams in gullies to raise local water tables and bring karez back to life. In such inhospitable terrain, these mysterious tunnels could be ready for a comeback. And ideas for reviving ancient *qanat*-like systems for taking water from wet mountains to dry lowlands are catching on in unlikely places. Such as Peru.

≋≋

Ancient South American societies, like their old-world counterparts, had many ingenious devices for managing scarce water in inhospitable terrain. A few of them are being revived. One is a pre-Inca technology in the mountains of the Andes that engineers believe could provide a cheap way of keeping faucets flowing in the drought-plagued Peruvian capital of Lima.

Lima is the world's second-largest desert city, second only to Cairo. It relies for water on rivers that flow out of the Andes. But those rivers diminish to a trickle during the long dry season, leaving the population of almost nine million with intermittent water supplies. Many of the city's poor rely on expensive supplies brought in by tanker.[13] But in 2015, the city's water utility announced plans to revive a system of ancient stone canals, known locally as *amunas*, that were built in the Andes by the Wari culture between AD 500 and 1000.[14]

The *amunas* were ingenious. They captured water from rivers in the Andean mountains during the rainy season and channeled it downhill for 1,000 feet, to where it could infiltrate rocks that fed springs. Regularly recharged with water brought from the mountains, those springs increased their flow enough to maintain a number of local rivers through the deserts of lowland Peru during the dry season.

The Wari canals fell into disrepair long ago and had been largely forgotten. Some still carried some water, but in most places the water dissipated and no longer replenished aquifers. As a result, the rivers in the deserts have dried up in the dry season, but regularly flood in the wet season. Lima has water shortages for seven months of the year but suffers regular floods and landslides during the other five months, when the Rimac, Chillon, and Lurin Rivers, which pass through the city on their way to the Pacific Ocean, are in full flow.

Hydrologists such as Bert De Bièvre of Condesan, a Lima-based NGO behind the revival plan, say repairing the canals with cement would allow them to resume their original purpose. Once again they could convey floodwaters from the mountain rivers for storage beneath the land, ready to top up the rivers in the dry season. "We have been injecting ink into the canal water to see where it s," De Bièvre told me. He is confident that the project could revive at *amunas*, mostly in the catchment of the Chillon River.

t should be enough to increase dry-season water supplies to Lima by acre-feet, reducing its current water deficit. It could be achieved at a s than a hundredth that of the city's new desalination plant. The city's utility, Sedepal, decided to fund the $23 million plan by taking 1 per- om its water charges.[15] The Wari people, I can't help thinking, would be of the continuing value of their handiwork.

X

When the rivers run dry . . .

we go looking for new water

32

TAKING THE WATER TO THE PEOPLE

IT WAS FOR SEVERAL YEARS the world's largest civil engineering project and aimed to remake the natural hydrology of the world's most populous nation. Chinese leaders saw it as a logical extension of the grand schemes of great societies down the ages. Ancient Mesopotamia and Egypt both harnessed their rivers to feed their people. The Romans were famous for their aqueducts. Persian empires were built around laboriously excavated tunnels that delivered water from deep underground. The Wari had their *amunas*. But even at their greatest extent, these ancient works were mild modifications to natural drainage patterns. Today's engineers have bigger ambitions, diverting entire rivers onto distant plains. China's scheme was the most ambitious ever. It would, the world was told, enable the Middle Kingdom to continue feeding itself as it had done for thousands of years. Maybe.

The South-North Water Transfer Project diverted part of the flow of the Yangtze, the world's fourth-biggest river, to replenish the Yellow River and supply the tens of millions of people in megacities that rely on it in northern China. The price tag was put at $60 billion, more than twice the cost of even Gadaffi's fantasy-world Great Man-Made River. It would eventually deliver twenty times more water than Gaddafi's pipe dream.

The project is actually three separate diversions. Two of them are finished. The first enlarged the Danjiangkou Reservoir on the Han River, a major tributary of the Yangtze, and takes 7.7 million acre-feet of its water north each year, enough for 50 million people. The reservoir was already Asia's widest artificial expanse of water. Its enlargement flooded another 140 square miles and, in a densely populated part of the country, displaced around 350,000 people. The canal north from the reservoir is 200 feet wide and as long as France. As it crosses China's crowded plains, it spans 500 roads and 120 rail lines, and tunnels beneath the Yellow River through a giant inverted siphon. It was inaugurated in late 2014.[1]

The second diversion, which also began deliveries in 2014, takes water from near the Yangtze's mouth to the megacity of Tianjin, which has suffered chronic water shortages since the 1990s. Part of the route uses the twenty-five-hundred-year-old Grand Canal, which was once the world's largest artificial river and the first to have lock gates. The third, western route is the most ambitious. It will take water from the Yangtze headwaters amid the glaciers of Tibet and push it through tunnels up to 65 miles long into the headwaters of the Yellow River. There is no firm route yet, but several Yangtze tributaries will be drained, and there is talk of building the world's highest dam. Altogether the plan is to send some 36 million acre-feet of water a year from the Yangtze to northern China.

On the other side of the Himalayas, in India, the South-North scheme is being seen as a template for an even larger project in the world's second-most-populous nation. The River Interlinking Project is intended to capture the great monsoon rivers of northern India, such as the Ganges and the Brahmaputra, and send their water south and west to the parched lands where the droughts are worst and the underground water reserves are sinking fastest.

The Indian scheme is far more complex than the Chinese plan. It would build dozens of large dams and canals to link fourteen northern rivers flowing out of the Himalayas. It would pump their waters south along a thousand miles of aqueducts and tunnels and through three hundred reservoirs to fill a second network, linking the seventeen major rivers of the country's arid south. These rivers include the Godavari, the Krishna, and the Cauvery, each of which has been diminished by heavy overabstraction for irrigation. The transfers would involve moving 38 million acre-feet of water a year, very similar in scale to China's scheme. That water could, say proponents, irrigate up to 85,000 acres of farmland and generate 34,000 megawatts of electricity. With so many more rivers and links, the price tag would be two to three times higher than China's project, with official estimates ranging from $112 billion to $200 billion, or around 40 percent of the country's GDP. But at least it could be done in many more small steps.

The River Interlinking Project has a long history. The legendary nineteenth-century British engineer Sir Arthur Cotton first conceived of a network of canals linking rivers as a means of improving the country's transportation. In the end, the British decided to build railroads instead, but the idea never quite died. For a while it was poetically described as providing India with a "garland of canals." However, it took a drought across India in the summer of 2002 to bring the idea back to the fore. The specter of famine returning to India after more than a decade in which the country had had food to spare was a profound shock. It has not receded. With water tables tumbling and the country's population predicted to increase by half a billion within the next half century, overtaking China, there has been a growing sense that a hydrological holocaust might not be far away in India. Many Indian politicians believe the only answer is to replumb the nation. The project gained the backing of Prime Minister Narendra Modi after his election in 2014.[2]

The first small phase, a canal joining the Godavari and Krishna Rivers in the east of the country, was completed in 2015.[3] Plans for more small links in the chain came thick and fast after that.[4] "We are going ahead with five links now," said Modi's water minister Uma Bharti.[5] It is easy to see the political imperative for this. India is beset by interstate rivalries over water that constantly threaten to boil over. In northern India, Punjab and Hiryana are at loggerheads over the Ravi-Beas River. Hiryana claims that since 1976 it has been entitled to 3.5 million acre-feet of water from the river, and it is demanding that Punjab complete a canal to supply that water. Punjab argues that it has no surplus water to share.[6]

Meanwhile in southern India, the states of Karnataka and Tamil Nadu have been feuding over the Cauvery River for decades. The argument began in 1927, when upstream Karnataka built the Krishnaraja Sagar Dam, creating an 80-mile-long reservoir on the river. Tamil Nadu responded with the Mettur Dam in the 1930s. Both states built extensive irrigation systems and encouraged their farmers to grow water-intensive cash crops such as rice and sugarcane, in efforts to cement their claims to the bulk of the river's water.

Initially this was just economic one-upmanship. But the stakes have risen. Today both states have populations as large as those of major European countries. On top of escalating rural demand, cities such as Bangalore, the heart of India's Silicon Valley in Karnataka, are demanding ever more water for themselves. Estimated demand from the two states has now reached 20 million acre-feet a year. Average flow on the Cauvery is only 16 million acre-feet. Inevitably, the upstream state, Karnataka, is in the driver's seat. Many years it refuses to deliver downstream the water that Tamil Nadu expects and demands. In September 2016, there were violent protests in Bangalore, after the Supreme Court ordered the state to release water down the river to its neighbor. Karnataka had refused on the grounds that a poor monsoon had left thousands of its own irrigation tanks dry.[7] But the Supreme Court forced it to release 12,000 cubic feet per second for five days, or a quarter of the then flow. No long-term solution has yet emerged.

Whatever the political imperatives, there are many doubts about whether the grand interlinking plan is either doable or sensible. Indian critics say that a third of the power promised from hydroelectric turbines to be installed as part of the scheme would be soaked up in pumping water around the network of canals and tunnels. Another concern is that the project would not just distribute water around the country but would also distribute the rampant pollution in northern rivers. And one estimate holds that an area of land the size of Cyprus would have to be flooded for the various reservoirs, canals, and structures, leaving three million people homeless.

Bangladesh objects too. It is downstream of India on the two biggest rivers to be exploited for the transfers, the Ganges and the Brahmaputra. Bangladesh relies on those rivers to irrigate its crops in the long dry season and to recharge underground aquifers. It has long complained that India's Farakka Barrage on the Ganges close to the border diverts valuable dry-season water away from Bangladesh and into Indian irrigation canals. In 1996, India promised not to reduce the flow further. Yet now, Bangladeshis complain, it is talking about doing precisely that.

Environmental economists concluded that the project's plans for the Brahmaputra would cause "a 10 to 20 percent reduction in the river's flow [that] could dry out great areas [of Bangladesh] for much of the year." Not only that but, without the flow of freshwater, salt from the Bay of Bengal would invade the large delta regions of southern Bangladesh. The result would be "an environmental catastrophe," say Edward Barbier of Colorado State University and Anik Bhaduri of the International Water Management Institute in Delhi, India.[8]

Whatever the drawbacks of such schemes, politicians and engineers around the world are touting their own grand plans for moving water between river basins. Iran plans a 285-mile underground pipeline to take desalinated seawater from the Caspian Sea in the north of the country through the Alborz Mountains to arid heartlands such as Qom, Khorasan, and Isfahan. Isfahan was once watered by the river Zayanderud, but the river has been running dry since 2012, thanks to irrigation farther up the valley.[9]

There is also a surprisingly strong political push behind a megaproject to harness the flow of the Congo, the world's second-biggest river, to irrigate the fringes of the Sahara and refill Lake Chad. In 2002, several governments signed an agreement on sharing the waters of the Congo. They wanted to extract water from one of the Congo's major arms, the river Ubangi, at Palambo, a sleepy river port in the rainforest 60 miles north of the capital of the Central African Republic, Bangui. The water would be poured into a pipeline and diverted over low hills into the nearby headwaters of the river Chari, from where it would flow downstream to Lake Chad 1,500 miles to the north. The project would be "an opportunity to rebuild the ecosystem, rehabilitate the lake, reconstitute its biodiversity, and safeguard its people," said Adamou Namata, the water minister in Niger and the chairman of the Lake Chad Basin Commission. "If nothing is done, the lake will disappear."

A feasibility study paid for largely by the Nigerian government recommended, in 2011, a transfer of 2.8 million acre-feet per year, equivalent to a fifth of the total flow of the Colorado River. The infrastructure would cost $7 billion, and the transfer would be enough to raise water levels in Lake Chad by about 3 feet and restore much of its former surface area. Half a decade on, with millions of refugees fleeing the dried-up lake region, they were still talking.

In January 2017, Nigeria and Chad again called on donors to fund their plan to "recharge" the lake, this time with an estimated cost of $50 billion.[10] But a study by the German development agency GIZ concluded that it would cause as much social and ecological chaos as the lake's drying out had. "Hundreds of thousands of people [will have] to hurriedly clear the areas that will be resubmerged," said UNESCO's Njike Ngaha.[11]

33

SINGAPORE

Sewage on Tap

IN SINGAPORE, the government likes handing out a new brand of bottled water—NEWater. It's probably the only bottled water in the world guaranteed to contain sewage. Albeit cleaned-up sewage. This may be the future. Because cities around the world see high-tech Singapore as an urban water pioneer—the testing ground for green technology that will keep the faucets flowing in a world of droughts, climate change, and rising city populations.

Singapore is a city state, an urbanized island of five million people that has traditionally got its water by pipeline from reservoirs in Malaysia, a neighbor from which it declared independence half a century ago. It wants to end its lingering hydrological dependence and cut the pipelines before the current supply contract runs out in 2061.[1] Its pursuit of water independence has turned Singapore into a world leader in optimizing water management.

For Yap Kheng Guan of Singapore's Public Utilities Board, which is in charge of filling the city's faucets, the route to self-sufficiency lies in "closing the loop" by recycling water, or "connecting toilet to tap." Singapore has built five treatment plants that collect sewage and clean it to such a high standard that it can be recycled back into the water supply. Sophisticated membranes remove solids, and ultraviolet light disinfects what makes it through the membranes. The end product "is cleaner than regular tap water," Yap says.

While most of the recycled sewage goes to local industries that need very pure water, such as microchip manufacturing and pharmaceuticals, some is added to the city's drinking-water reservoirs. To head off public complaint, they mix it with regular water from the city's rains and treated runoff from its streets. They call the resulting fluid, which is about 2 percent treated sewage, NEWater. Besides filling an estimated twenty million water bottles so far, NE-Water meets 30 percent of Singapore's domestic water needs.

Water is never wasted in Singapore. The country has also engineered its cityscape so well that two-thirds of its land surface funnels rainfall into urban reservoirs. When I visited the city, the rains were beating down as a storm came in from Sumatra. But barely a drop escaped Yap's water catchers. Altogether, urban drains supplied a third of the city's water. Even the rain falling on runways at Changi Airport fed faucets in the terminals. "We are saving every drop of water that we can; we are closing the hydrological loop," said Yap. Of course, runoff from city streets isn't clean enough to drink without treatment. So to keep cleanup costs down, Singapore is sanitizing the catchment too. Livestock farms have been banished, industrial discharges purged of toxins, and the public told to take their trash home.

In most cities around the world, leaky water mains are a big problem. Many lose a quarter of their water this way. Some—London, Nairobi, and Shanghai included—squander between a third and a half of all the water put into the mains. Yet Singapore has cut leaks to what is probably a world-record low of around 5 percent. It helps that the city has few nineteenth-century water mains. But the clever bit is the way it has installed smart valves and sensors that control water pressure throughout the system and automatically "listen" for leaks. Bursts that in many cities would go undetected for months or years can be fixed in Singapore within hours.

Many cities spend billions of dollars on plumbing to rush floodwaters off streets and into oceans. Then they spend billions more transporting more water from distant hills to keep the faucets flowing. Singapore considers that bad hydrology. It recycles as much as it can. The truth, however, is that by turning only its sewage effluent back into drinking water, Singapore is missing a trick. For the effluent contains something else of value—the sewage itself. For Singapore, as in most cities, it is a waste product. But all that feces and urine is just what some people would like to get their hands on. It is a potentially valuable resource, especially as agricultural fertilizer. Our misuse of human excrement must count as one of the modern world's worst, but least discussed, resource failures. If you'll excuse the phrase, it is shit economics. And pretty lousy ecology too.

Sewage is turning rivers into cesspools, destroying soils, harming human health, damaging food supplies, and wrecking water supplies. This is happening because we refuse to see that the stinking, lumpy pathogen-rich effluent pouring from sewer pipes and festering in latrines is not waste: it is free water for irrigation or other uses, and rich in nitrogen, phosphorus, and lots of other things that growing plants need. Flushing sewage into rivers is not just an environmental catastrophe for the rivers; it is also a ludicrous waste of nutrients that could be helping to feed the world.

Consider what you excrete. In a typical year, your body will relieve itself of some 130 gallons of urine and 110 pounds of feces. That is sufficient to fertilize plants that would produce almost 500 pounds of cereals, says Christine Werner of the German development agency GIZ. The world needs a strong dose of what English nannies would call potty training. For the good of both the environment and agriculture, we need to put our poo in the right place—onto fields. So let's cheer the work of the honey-suckers of Bangalore, India's most high-tech city. They are the pioneers we need.

"Honey-suckers" is the name locals in Bangalore give to the makeshift trucks, equipped with tanks and suction pumps, that empty the septic tanks and pit latrines where more than half of Bangalore's nine million inhabitants deposit their daily waste. Until recently, the trucks conveyed this effluent to local streams and lakes, where it added to a growing pollution crisis. But now the drivers head instead for farms around the city, where the sewage is decanted into trenches and used to fertilize coconut, banana trees, and other crops.

The farmers pay good money for human waste because it produces bumper crops. For them, it is as sweet as honey. So the small-time owners of the three hundred or so honey-suckers that ply Bangalore's streets get paid twice: once by the householders and apartment-block owners eager to have their overflowing tanks and pit latrines emptied, and a second time by the farmers. With hard work on the honey run, the drivers can recoup the cost of their trucks within six months, according to Vishwanath Srikantaiah of Biome Systems, a Bangalore-based green consultancy that has investigated the practice.

Farmers don't always pour the sewage straight onto their land. Typically, they put it into drying pits. That kills pathogens and concentrates the nutrients. It also means the dry effluent can be dug into the soil more easily. But during the dry season, when plants need extra water, the farmers will pour the still-liquid sewage into channels beside their crops. It is then providing irrigation as well as nutrients. Vishwanath says that septic tanks emptied by honey-suckers offer not only a cheap alternative to the expensive construction of sewers down every street in the city, but a superior solution. What's not to like?

A billion or more urban dwellers around the world live in homes that are not connected to sewer systems. In India alone, 160 million households defecate into septic tanks and pit latrines. These waste repositories are emptied by a vast but largely secret industry. Despite laws banning the practice, an estimated one million Indians from lower castes, mostly women and girls, are paid a pittance to scrape the shit from the nation's one hundred million or more tanks and latrines. Usually their tools are nothing more than a shovel and bucket. They dump the contents in nearby drains or on waste ground.[2]

This is a human scandal, and also an environmental one. Recycling our waste onto fields would increase food output and make life a lot easier for poor farmers, who often cannot afford chemical fertilizer. A typical family in Niger, one of the world's poorest countries, annually excretes nutrients equivalent to 200 pounds of chemical fertilizer, according to Linus Dagerskog of the Stockholm Environment Institute. That fertilizer would have a market value equivalent to a quarter of a typical rural income.[3]

Scale that up and the world's population excretes almost 80 million tons of nutrients annually. Applied to fields, this could theoretically replace almost 40 percent of the nutrients in chemical fertilizers used by the world's farmers. So maybe it is time to rethink how the world manages sewage, not just in places without sewer systems but in those with them too.

Around the world, city authorities are tasked with removing human waste. The fast-growing cities of the developing world are trying to deal with their waste in the way most industrialized countries do, by connecting every building to sewer networks that run down every street. They strive to attach those sewer networks to treatment works that remove solids and other potentially dangerous contaminants before discharging the partially cleaned effluent into the nearest river. Mostly they fail. Few of the world's megacities—and even fewer of the thousands of medium-sized urban areas—have fully functioning sewer networks. Around half of all the world's urban homes are not connected to sewers. Even when they are, the sewage rarely heads for treatment works. Only around one-tenth of urban sewage is treated. Most of the remainder is dumped directly into the natural environment, usually into rivers, where it turns thousands of miles of waterways into lifeless open sewers and creates dead zones that now cover 10,000 square miles of ocean.

This is not just a problem of sewer systems. It is a problem of water supply too. Sewers only function if they connect to domestic flushing toilets. But those too are rare, because the water required to flush them may be in short supply. As much as half the water in a typical Western home is used for flushing. In many places, that water could be much better used for drinking or irrigation.

Growing cities too often take water from farmers who need it to irrigate crops and feed growing populations.

So if "flush, collect, and treat" sewage systems are a waste of valuable re-sources and a recipe for pollution, what is the alternative? The answer may be a return to the past. Before the invention a century ago of the Haber-Bosch chemical process for converting nitrogen from the air into nitrates, human sewage was the fertilizer of choice. Traditionally, it was collected in the dead of night to avoid offending people's sensibilities—hence the term "night soil"—and was used to grow vegetables and other crops on the outskirts of cities. A "sewage farm" was just that.

Campaigns to improve public health and the introduction of flush toilets meant that the practice grew obsolete in most cities in the industrialized world. But even where sewer systems developed, farmers still sometimes competed for the network's outpourings. In a few places, this has persisted. Since the 1890s, most of the sewage from Mexico City has been piped untreated to the fields of Tula Valley beyond the city's northern suburbs. Today, the excrement from the megacity's twenty-one million people continues to fertilize more than 250,000 acres. It has made Tula Valley farmers wealthy. The remains of the city's digested beans, tortillas, and chili peppers double their yields of corn and almost triple the rentable value of farms in the valley, says Blanca Jiménez of the Mexican Academy of Sciences.[4]

Sewage is especially valued in dry regions where farmers need the guaran-teed year-round irrigation as much as the nutrient supply. Chris Scott of the International Water Management Institute has come to the startling conclu-sion that perhaps a tenth of all the world's irrigated crops—everything from rice and wheat to lettuces, tomatoes, mangoes, and coconuts—are watered by the untreated stuff disgorged from the end of sewer pipes. The practice is most frequent on the fringes of the developing world's great cities. In such places, clean water can be in desperately short supply in the dry season, while sewer pipes extending out of the cities keep gushing their contents onto the nearest open land all year round. It's no wonder farmers compete to grab the stuff for their fields.[5]

In Hyderabad, the Indian city where Scott once worked, "pretty much 100 percent of the crops grown around the city rely on sewage. There is no other water available," he says. In Pakistan, sewage is used to grow perhaps a quar-ter of the country's vegetables. Farmers using sewage for irrigation typically earn $300 to $600 more annually than those without the benefit. Such benefits explain why, in parts of Pakistan, it costs twice as much to buy fields watered by sewage pipes as neighboring fields irrigated with clean water. And why, in the Indian state of Gujarat, farmers compete for the sewage at annual auctions, preferring it to freshwater irrigation. Some of the sewage is delivered by truck. Some is simply discharged onto fields from the end of sewer systems.[6]

The biggest argument against agricultural recycling of sewage is that it carries disease that threatens either farmworkers or consumers of their crops. While urine is largely pathogen-free, fecal matter is rich in viruses, bacteria, and worms. The World Health Organization estimates that there are more than two million deaths a year worldwide from diarrhea and other diseases associated with human waste, though nobody knows how much of this is connected with using sewage in farming.

Some researchers are sanguine. They say that soils quickly filter and clean bacteria found in feces.[7] Other studies reinforce the doubts. A rare Indian investigation looked at twenty-two farming villages near the Musi River, which these days is little more than a sewer for Hyderabad. It found that almost half of households irrigating their fields with the sewage reported fever, headaches, and skin and stomach problems during the previous year—twice the rate in a control village that used clean water for irrigation. The highest disease rates were among women who weeded the fields.[8]

The instant reaction is to ban sewage farming. The trouble is that it is often banned already, with little effect in the real world, where farmers value free nutrients and irrigation. A more practical approach might be to improve hygiene by encouraging safe practices for handling fecal matter.[9]

The truth is that, in an era where recycling and "closing the loop" on key resources is the new normal, the days of "flush and forget" may have to come to an end. Even in the developed world. We should be looking for ways to recycle our urine and feces, in the same way we recycle scarce metals. Not in the raw, but after treatment to make it safe. There are pioneers. Israel now recycles more than 80 percent of its treated sewage onto fields. It is the country's default source of irrigation water. Neighboring Jordan recovers 60 percent.[10]

Mexico recycles enough treated wastewater to irrigate around 600,000 acres. On the Rio Grande, I visited a giant new state-of-the-art sewage treatment plant at Juarez, El Paso's twin city. It treated half the city's sewage and delivered enough effluent down a canal to irrigate 75,000 acres. It was virtually the only source of water for crops downstream on the Mexican side of the river. El Paso has long done the same thing for watering its parks, but now plans to irrigate crops too.

Orange County in Southern California has gone further. It pours treated sewage into rocks beneath the county. There, the geology filters out the nasties before the county pumps the water up again to fill the faucets of more than two million residents. They do much the same at the ancient cathedral city of Winchester in England. Overlooking the city is St. Catherine's Hill, a beauty spot with a dirty secret. Just beyond the hill's ancient fort, Winchester pours its sewage into the hillside, from where the local water company eventually supplies it to faucets across South Hampshire. It's quite safe. The chalk cleans the sewage well.

I have drunk cleaned-up sewage all my life. I live in London. Our water has typically been drunk several times by people living in towns upstream on the river Thames. In Swindon, Reading, Maidenhead, and elsewhere, they abstract, drink, excrete, and return the river's water before we get our chance to take our turn. The advanced water-treatment technologies in use at every step make it as safe as water anywhere, despite its unappetizing recent history. Our sewage effluent spends time in a river rather than a pipe, before being recycled into the water supply system. But otherwise things are little different from Singapore. As rivers run dry and groundwaters diminish, more and more cities round the world are coming up against hard limits on water supply, while demand continues to grow. From Cape Town in South Africa, scene of a water crisis in 2018, to Bangalore in India, from Mexico City to innumerable cities across China, the idea of single-use water must come to an end. Recycling is urgently required. Places like Singapore and London may become the norm. Our water may increasingly come not from the natural water cycle but from our own industrialized version.

34

OUT OF THIN AIR

THE DEW HUNG HEAVY over the Sussex Downs near the ancient Roman city of Chichester. Peering through the swirling mist of a November morning, I spied a pond. It appeared almost magical, like an oasis in the desert. Though the air was damp, this patch of water was far from any spring or stream and too high to gather water from the surrounding pastures. This was also an unfailing pond. It had not emptied for decades, not even during the notorious long drought of the mid-1970s, which had killed streams, dried up springs, and turned fields hereabouts almost to desert. I was looking at a dew pond. Its secret was to capture unseen moisture from the air in the hills. Though largely forgotten today, dew ponds were once essential to the great sheep pastures that covered these chalk hills. They are a hidden mystery of the English landscape, only now being recognized and revived, out of nostalgia and a desire to provide water for wildlife.[1]

For centuries, dew ponds flourished across southern England. Gilbert White, a famous eighteenth-century chronicler of the English countryside, wrote of a dew pond high above his home in Selborne in Hampshire. It was, he wrote, "never above 3 feet deep in the middle, and not more than 30 feet in diameter, and contained perhaps not more than two or three hogsheads of water, yet it is never known to fail, though it affords drink to 300 or 400 sheep and for at least 20 head of large cattle besides." While the valleys dried up in summer, he said, the pond on the hilltop always kept its water.[2]

Naturalists often imagine that dew ponds are prehistoric features. That is understandable: many of them still survive near ancient monuments. On the South Downs, I saw dew ponds near Iron Age forts at Cissbury Ring and Chanctonbury and close to the Celtic temples of Lancing Ring. In his story "Weland's Sword," Rudyard Kipling talks of a time long ago, "before the Flint Men made the dewpond under Chanctonbury Ring." The celebrated war photographer Don McCullin hinted at the same antiquity when he opened and closed a book of his best photographs with ethereal shots of dew ponds close to Glastonbury Tor in Somerset, site of the first Christian church in England and reputed home of Camelot and King Arthur.

Most dew ponds are more recent. They were dug to provide water for the large flocks of sheep that grazed on the downs in the eighteenth and nineteenth centuries, but were abandoned when the decline in sheep and the arrival of piped water in the twentieth century made them largely obsolete. One of the last diggers was a Mr. Smith from the Chiltern Hills. He advertised his craft into the 1930s, providing a phone number for free estimates for anyone wanting to employ him in "a secret process handed down from father to son over 250 years." His ponds were "guaranteed to condense and retain beautiful clear water without the aid of pumping etc."[3]

So what was the "secret process"? On chalk hills, water quickly slips underground through the porous rock, so the first requirement is a waterproof lining. Dew-pond diggers used local clay, "puddled" to make it watertight by driving teams of oxen through it. But that doesn't explain where they got their water. There is a ticklish mystery here. Some water comes from rain, to be sure. However, the dew ponds' real trick is to capture moisture from the air, catching the tiny water droplets in passing clouds and fog and encouraging more water vapor to condense on the ground in the form of dew.

Here, the hilltop location is the key, because as air rises it cools. Cold air holds less moisture, so the rising air creates water droplets that form clouds or fog. At night, as the hill air cools further, more water condenses and dew forms. Martin Snow, a local historian, has been mapping dew ponds on the Sussex Downs for years. He says that many of them are cleverly sited at the heads of dry valleys that channel mists and moist air uphill. The old sheep farmers, he reasons, noticed that fog lingered at the heads of these valleys and the grass was often moist. "Creating ponds would have been a natural way to catch that water," he told me.

The secret process guarded by the pond diggers was more than a matter of siting, however. Their trick was to create a pond that captured the maximum moisture by providing as cool a surface as possible on which dew would condense. They did this by putting straw beneath the clay and stones on top. The straw insulated the clay, allowing it to cool more than the soil beneath at night. The stones, which shed heat quickly at night, lowered the temperature further. Once a successful dew pond was created, it would, in effect, generate its own water from the air.

Many dew ponds have been plowed up in recent decades as pastures have given way to arable farming. Others are overgrown. On the hills above Worthing, residents still remember how shepherds would halt their flocks for a drink at the Tolmare Farm dew pond on the way to the Findon Sheep Fair. Today that pond is empty, with a radio tower through its center. At Cissbury Ring, military tanks have pierced the pond's clay crust. But a surprising number of these ponds just keep on catching moisture from the air, and dew ponds have benefited from recent national efforts to restore and dig ponds of all kinds.[4]

The National Trust renovated one on Bignor Hill in an ancient landscape of Bronze Age barrows crossed by Neolithic causeways and a Roman road. The Sussex Downs Conservation Board dug a new one at Sussex University in 2004. Sadly, says Snow, "most of the renovators are more concerned with providing water for wildlife than with re-creating dew ponds for their original purpose." The Bignor dew pond is full of reeds and yellow flag and the pink-flowered amphibious bistort. A worthy aim, of course. However, it is fenced off from the sheep that it was dug to provide for.

Dew collection is not unique to the chalk hills of England. The otherwise waterless Croatian island of Vis in the Adriatic Sea has paved surfaces on hillsides behind a Franciscan church that collect moisture from hot, dry summer air and channel it into cisterns.[5] On the Canary Island of Lanzarote, few vacationers will have noticed that local farmers surround their crops with a mulch of volcanic gravel that condenses atmospheric moisture.

Lanzarote, an island off the west coast of Africa, was a tranquil place in the eighteenth century, ruled by Spanish priests and visited occasionally by ships making the transatlantic crossing. Farming was rudimentary and the living poor. The island had less rain than much of the Sahara Desert. Then came a series of massive volcanic eruptions that shook the island almost without a break from 1730 to 1735. A priest described how, at the height of the eruptions, "the earth suddenly opened . . . a gigantic mountain rose and sank back into its crater on the same day, covering the island with stones and ashes."[6]

It could have been the end for the island's inhabitants. Yet, far from obliterating its life, the eruptions provided the key to the island's future prosperity. The black volcanic stones noted by the priest launched a bizarre agricultural revolution. As the farmers returned to their fields and contemplated the task of removing the stones, they noticed something odd. In the areas covered by stones, their crops were bursting forth, whereas elsewhere they were not. It didn't take the farmers long to discover why. The black pumice-like stones had shaded the soil from the glare of the sun, reducing evaporation and keeping the soils damp. They were also porous and trapped moisture by capturing nighttime dew. We now know the stones—which the local people called *picon*—cut water loss from the often-parched fields by around 75 percent. New crops would grow in the soil, including fruit and vegetables, provided there was a layer of *picon* at the surface.

Soon the farmers were importing camels from Mauritania to haul stones to fields that had not benefited from the eruptions. Formerly useless hillsides were cleared of brush, cut into terraces, covered in *picon* and planted with vegetables. By 1776, an anonymous chronicler was recording a "prodigious mutation" of farming on the island. "Marvels abound, with the land being more fertile, becoming fruitful and bearing fruit two or three times a year. Like sponges, the *picon* soaks up the water, and the crops receive a delicate, gentle watering," he wrote. "Before the eruptions in 1730, the most the island produced was bread and beef; now on the strength of the picon, it produces grapevines, vegetables, maize, potatoes, pumpkins and other produce."[7]

Exports of surplus produce began. Grapes did particularly well. The farmers grew prickly pears for the cochineal insects that infested them. The insects are deep crimson inside, and the islanders scraped up the insects with a spoon and dried and ground them, before selling the vivid crimson remains as a dye. Cochineal harvesting was big business in Lanzarote in the nineteenth century, before the advent of synthetic dyes.

The stones revolutionized agriculture. Their water-catching abilities remain vital on an island with no permanent rivers, few underground water reserves, and rainfall averaging just 37 inches a year. Nowadays, most of Lanzarote's population make their living from the two million tourists who visit each year. But the black fields produce nearly half a million gallons of wine a year. Lanzarote cochineal is still used in everything from lipstick to strawberry milkshake and Campari.

Farmers reckon that fields need to be replenished every thirty to fifty years, as the stones become mixed into the soil. But there are still plenty to go around. "Virtually all cultivated fields in Lanzarote still have a layer of *picon* today," says David Riebold, a British forestry scientist on the island, who showed me round. "Without *picon*, you couldn't grow crops in most places."

The technology has some precedents. It seems the prehistoric inhabitants of Easter Island in the Pacific used mulches of volcanic stones to trap moisture, especially after they had destroyed their forests. Some farmers in arid western

China cover their fields with gravel from riverbeds to cut down evaporation. But like the dew ponds in England, *picon* remains a remarkably local hydro-logical phenomenon that goes largely unnoticed both by the world's agricul-turalists and by the millions of tourists taking the sun on this very unusual desert island.

≈≈

Dew ponds and stone mulches may seem like local curiosities, but they raise an interesting question. Could we be harvesting moisture from the air on a much bigger scale than we do? After all, the atmosphere contains around 10.5 billion acre-feet of water, six times more than all the world's rivers. At any one time, about 2 percent of this atmospheric moisture has condensed into water droplets in clouds ready to fall as rain. The remaining 98 percent is a tantalizing prize, and some outlandish machines have been proposed to capture it.

In the 1930s, a French meteorologist named Bernard Dubos wanted to build a chimney 2,000 feet high with a fountain at its base to create an updraft of humid air that would, he believed, saturate the air above and generate rain. The machine was never built. But meteorologists have from time to time ex-perimented with manipulating the electrical charges in clouds to encourage the formation of raindrops. In the 1940s, South Africa's chief meteorologist, Theodore Schumann, proposed an electric fence 150 feet high and 2 miles long on the top of Table Mountain outside Cape Town to condense millions of gal-lons of water a day from the atmosphere. It might have come in handy when the city's water ran out in 2018. In the mid-1990s, I came across a small Rus-sian unit doing something similar in eastern Jordan. If it worked, they never claimed success.[8]

Some have also suggested using sound to capture moisture from the air. On a cool, still night, the air can be so saturated with moisture that even mod-est air movements, such as sound waves, can produce raindrops. In the moun-tains of Yunan in southern China, villagers have a tradition of yelling loudly in the hope that it will stimulate rain. The louder they shout, it is said, the more likely it is to rain. This gives interesting scientific credence to the African tradi-tion of the rain dance.

More straightforward is the idea of harvesting water from fog. This was the brainchild of Bob Schemenauer, a Canadian cloud physicist, who first used it thirty years ago to provide water for a small run-down fishing town on the edge of Chile's Atacama Desert. Chungungo goes years without rain, yet fogs regularly roll in off the Pacific Ocean. Schemenauer erected what looked like a volleyball net for giants: seventy-five large sheets of plastic mesh suspended along a remote ridge above the town. Droplets in the fog coalesced on the mesh and drained into collecting bottles. The scheme was an immediate success in a town of 350 people who had previously depended on a water tanker driven from 50 miles away. Each sheet, measuring 40 feet by 10 feet, took 40 gallons of water a day. At one point in the late 1990s, the nets were delivering an average of 4,000 gallons to the town each day.

"Until we set up the water system, the town was dying," Schemenauer told me at the time. "But with a secure water supply, people began to return, and new houses were built." He anticipated expanding the system. There was plenty of room on the ridge. Five years later, he was wiser. The town had grown so much that the government decided to install a pipeline to a distant reservoir. After that, neither the local authority nor the townspeople maintained the sheets. The system is now defunct, the plastic sheets torn and abandoned.

Schemenauer moved on. He tried his fog-harvesting system in other arid but foggy lands, from the Caribbean island of Haiti to Namibia, Nepal to Yemen, and Guatemala to Eritrea. His most successful trial was at Mejia, in the coastal hills of southern Peru, just up the coast from Chungungo. He erected five acres of nets, funded by the European Union. But when the project finished and the foreign scientists left, the locals abandoned the nets.[9]

Others have tried the same trick. Back on Lanzarote, David Riebold had a plan to harvest the fogs that rolled across the island to bring back cloud forests on the mountaintops. In 2005, the town authorities erected eight modest fog-collecting devices on three of Lanzarote's mountains. When I visited the following year, Riebold was getting an average of half a gallon of water a day from each square meter of net, enough to nurture two seedlings. He figured that after two years of watering from the nets, saplings would be tall enough to collect the fog water themselves on their leaves. Then the nets could be moved to the next site to begin watering another set of saplings. His project fell foul of local politics for a number of years, but in 2017, Riebold told me it had been adopted by a new environmental chief at the island council. So it could happen yet.

Nature is a good fog-catcher, too, and may have some tips for would-be fog harvesters. On El Hierro, another Canary Island, people harvested fog droplets from the leaves of trees until a hundred years ago. Perhaps this practice originated from a report by Pliny the Elder, in Roman times, of a Holy Fountain Tree growing on the island. In Namibia, British zoologists have discovered a desert beetle that has evolved a bobbled upper surface to its body with a hexagonal pattern that is supremely efficient at capturing moisture from passing fogs. The peaks and troughs appear to push tiny droplets together to form larger droplets, which then roll off the insect's back and into its mouth. Andrew Parker of Oxford University rigged up a prototype fog-catching surface based on the beetle's design, which captured five times more water than Schemenauer's netting.[10] He patented the design, with the idea of making fog-collecting devices to be put on the roofs of buildings, or even campers' tents. Others have tried developing the same for industrial applications and self-filling water bottles.[11] These are niche markets, but maybe someone somewhere will develop a method of fog catching that works at scale.

35

SEEDING CLOUDS
AND DESALTING THE SEA

A MORE CONVENTIONAL approach to conjuring water from the air is seeding clouds. Scientists have been active in this business for more than half a century. Like the grit in an oyster, particles in the air provide a necessary nucleus around which water vapor can coalesce into raindrops. Nature does this mostly with tiny salt crystals from sea spray. The idea of artificial seeding is to encourage the process by spraying billions of tiny particles, usually silver iodide crystals, into clouds.

Cloud seeding is good politics in a drought. As the planes take off and spray the sky, authorities can at least say that they are doing something to fill faucets and irrigation canals. In recent years, it has been carried out in at least twenty-four countries. The Indian state of Karnataka is a cloud seeder. So are the African governments of Morocco and Burkina Faso. Ten American states, from North Dakota to Texas, have tried it. Idaho has invested millions of dollars in the practice. Most places deliver the particles from planes, but the Chinese blast crystals out of canons, and Iran recently revealed it planned to use drones.[1]

Successes have been claimed, but proof is hard to come by. That is understandable. Who can predict what would have happened without the seeding? Even so, practitioners are convinced. Idaho claims to have increased winter snowfall and successfully promoted spring runoff into the state's rivers. This in turn has increased hydroelectric power generation and reduced the utility's need to buy power from neighboring states, it says.[2] In Australia, cloud-seeding experiments have been conducted in New South Wales to produce snowfall and maintain ski runs in the Snowy Mountains. Operators said they had induced a rise in precipitation of somewhere between 7 and 14 percent. The United Arab Emirates, in 2015, invested half a million dollars in 186 cloud-seeding flights, which, they said, played a part in delivering record rainfall and was much cheaper than desalination of seawater.[3]

Israel claims the most successful and long-standing cloud-seeding operation. According to the country's rainmaking guru, Daniel Rosenfeld of Hebrew
University of Jerusalem, it works best in winter clouds rolling across the north
of the country, over the hills of Galilee. He claims that spraying typically raises
rainfall from clouds by 15 percent and recent programs have produced 40,000
acre-feet of water a year—adding a couple of percent to the country's overall
water supplies.

The scientific basis for these claims is often not clear.[4] But while cloud seed-
ers suffer from an inability to persuade skeptics of their success, they also suffer
from fears that they could be all too effective. For one thing, cloud seeding can
be an act of war. The United States secretly seeded clouds over Laos and North
Vietnam in the 1960s to waterlog the trails of human donkeys carrying supplies to guerrillas during the monsoon. Sometimes seeding has been accused
of causing unwanted floods. Most notoriously, Operation Cumulus, a secret
military rainmaking experiment over southern England, may have triggered
huge storms on Exmoor in the summer of 1952. A few hours after one spraying, storms set off landslides that killed thirty-five people in the Devon village
of Lynmouth. The truth of cause and effect can never be known, but according to documents unearthed years later in the Public Record Office, Operation
Cumulus was abruptly suspended shortly after the disaster. Perhaps officials
knew enough.[5]

There are other problems. Cloud seeding, if badly done, can stop the rain.
Too many crystals thrown into clouds will cause the moisture to form too
many droplets, none of which become big enough to fall as rain. And even
if seeding generates local rain, the question remains whether that will reduce
rainfall somewhere else. Jordan, one of the driest and most water-stressed
countries in the world, fears that Israeli cloud-seeding may sometimes empty
rain clouds blowing east from the Mediterranean before they cross the Jordan
Valley. Canada's prairie provinces have had similar fears about US operations
in North Dakota.

China spends tens of millions of dollars a year on blasting its clouds with
silver iodide from ground-based rocket launchers.[6] As well as making rain, it
has seeded clouds with the aim of preventing hail formation, damping down
forest fires, dispersing fog, and even lowering summer temperatures. But across
the plains of eastern China, there have been dark warnings of rain theft. In the
middle of a dry summer in 2004, Pingdingshan city sent up planes to spray
passing clouds. Some 4 inches of rain fell on the city. But 90 miles downwind,
Zhoukou city received only a fifth as much. Zhoukou's city fathers concluded

their neighbor had emptied the clouds of the rain that was rightfully theirs. Maybe so, but I was only a few hundred miles away at the time, and it was raining where I was too—without the benefit of seeding. So who knows?[7]

It is well known that air rising over mountains cools, and because cooler air can hold less water vapor, the result is often clouds and rain. That is why mountains are usually wetter than plains. So how about making mountains? Abu Dhabi recently commissioned a study from an American university into how feasible that might be for the Gulf state.[8] Don't hold your breath. However, there might be more potential in another water-making scheme in the same emirate. This is the latest incarnation of an old fantasy to provide water by towing icebergs.[9] According to private-sector promoters of the project, an Antarctic iceberg that began its journey with 20 billion gallons of frozen water might, on arrival in the Gulf, still have a twentieth of its former volume, yielding enough water for a million people's domestic use for five years.[10] One day we might find out if that calculation is correct.

If conjuring water from the air or towing icebergs from the Poles remain high-risk enterprises, maybe desalination of seawater is the answer to the world's water woes. It seems like the commensense solution. After all, distilling seawater by boiling it and collecting the water vapor is an age-old activity.

Modern distillation technology was developed by the US Navy for operations on remote Pacific islands during the Second World War. Following that, large-scale distillation for public supply took off in the water-poor Gulf States, where they have plenty of oil to provide the large amounts of energy required. Today, global desalination capacity is about 70,000 acre-feet a day—roughly 1 percent of the global domestic water supply. Much of the global capacity is still in the Gulf, but Israel is now also a major producer.[11] Islands where summer tourists have overwhelmed local water supplies have until recently made up most of the rest. In Mediterranean Europe, Malta gets two-thirds of its drinking water from desalination. Greek islands like Mykonos have been doing the same for years, as have Caribbean islands such as Bermuda, the Caymans, Antigua, and the Virgin Islands.

Most of the world's total desalination capacity still uses distillation, but since the 1970s, an alternative technology has grown in popularity. Reverse osmosis forces water repeatedly through a membrane that filters out the larger salt molecules and lets clean water through. It too requires large amounts of energy. Tampa Bay, Florida, and San Diego, California, have taken the plunge with reverse osmosis, as has Perth in Australia. Spain built a string of reverse-osmosis plants after abandoning plans to pump water cross-country from the wet north to relieve the arid south.

Until recently, it cost ten dollars or more to produce a thousand gallons of unsalty water—up to a hundred times more than many conventional water supplies. That is why a desalination plant in Santa Barbara, California, sat mothballed for two decades before it was brought out of retirement at the height of drought there in 2015.[12] Costs for reverse osmosis in particular are coming down, and more cities are buying into the technology. Israel now claims to desalinate for around $2.50 per thousand gallons at its reverse-osmosis plants. That is around a third of the production cost for distillation in Saudi Arabia. Advances in membrane technology could cut costs further one day. Membranes made of graphene—a single layer of carbon atoms arranged in a lattice—are one possibility.[13]

Such prices are encouraging cities in cooler and wetter climates to join the reverse-osmosis revolution. China has a large plant in Tianjin, the country's third-largest city, where water shortages have been endemic for years. Even more surprising, Britain's Thames Water in 2010 completed a $400 million reverse-osmosis plant on the Thames in East London to process brackish estuary water during droughts. It decided the plant was cheaper than building a backup reservoir. Should it ever be required, the plant is able to meet the domestic needs of almost a million people.

The recent boom in desalination is beginning to alarm environmentalists. One problem is what to do with the salt extracted from the seawater. It emerges as a vast stream of concentrated brine. Most plants, naturally enough, dump it back into the sea. But this salty wastewater also contains the products of corrosion during the desalination process, as well as chemicals added to reduce both the corrosion and the buildup of scale in the plants. It is damaging for marine wildlife.[14]

Maybe this pollution can be fixed technically one day. But what can't be fixed is the huge energy demand of desalination. Even the most efficient reverse-osmosis plants require approaching 16 kilowatt-hours of electricity for every thousand gallons of water they produce. And the energy needed to pump the water uphill from the coast to the people who want it might easily double that. So while desalination could conceivably become a viable source of drinking water in coastal regions around the world in the coming decades, it would be at the expense of an extra push for climate change. It is also hard to see desalination ever penetrating the agricultural market, where the majority of the world's water is currently used. It is simply far too expensive. At the end of the day, desalination seems like an expensive high-tech solution to a global water problem that is overwhelmingly caused by wasteful use. Like the enormous engineering projects for shifting water around the planet, it is a supply-side solution to a demand-side problem.

XI

≈≈≈

When the rivers run dry . . .

we should go with the flow

36

EUROPE

Learning to Love the Floods

FROM THE INDUSTRIAL CITIES of Britain to the forests of Sweden, from the plains of Spain to the shores of the Black Sea, Europe is attempting to restore its rivers to something like their natural glory. The most densely populated continent on earth is finding space for nature to return along its riverbanks. The restoration is far from perfect. Fully functioning river floodplains cannot be restored when that would flood cities. But Europe's fluvial highways are becoming a test bed for realizing conservation biologist Edward O. Wilson's dream that the twenty-first century should be "the era of restoration in ecology."[1]

The political imperative is strong. The European Union Water Framework Directive, agreed to in 2000, requires that all rivers be returned to a "good status." That phrase has not been defined, but the direction of travel is clear. Water engineers in Europe have been trying to cleanse rivers of pollution for half a century, but they are now intent on restoring them to nearer their natural wild state. The belief is that more natural rivers will not just be good for ecology, they will be better for humans, too—less vulnerable to both floods and droughts, and better able to absorb pollution.

The change in fluvial fortunes has been dramatic. Two decades ago, I visited France's longest river, the Loire. I went to interview activists from all over Europe, who camped out on its banks near the town of Le Puy-en-Velay in southern central France through long winters to prevent construction there of a dam, the Serre de la Fare. It had seemed to me like a doomed battle. But on the final day of my visit, a dramatic victory was declared when courts sided with the activists. The dam was never built. Since then, rather than building dams on the Loire and its tributaries, engineers have been tearing them down. Dams such as the Maisons-Rouges, which had blocked salmon and eel migrations on the Vienne, are now gone. The engineers are re-creating what the environmentalists demanded—a "Loire Vivante," a free Loire.[2]

The mood has spread. Denmark's largest river, the Skjern, is getting back some of the marshlands at its mouth, after engineers reinstated meanders and lowered artificial banks to allow seasonal flooding. Arable fields have been returned to wet meadows. One of Spain's largest rivers, the Duero, is being cleared of dams and other artificial obstacles. Britain has promised to restore some 1,000 miles of rivers. Of twenty-seven hundred projects listed in its National River Restoration Inventory, more than half have been completed.[3]

River restoration means different things in different places. In some urban areas, it can mean simply taking a forgotten stream out of a sewer pipe or concrete culvert. In Northwest England, England's Environment Agency, a government body, is doing this for the Irwell and Medlock, both of which once ran lifelessly through Manchester, once the world's first great industrial city. Progress is slower in London. Take a trip on the city's Underground to Sloane Square station and you will find, hanging above the platform, a large green metal conduit that carries what remains of the river Westbourne on its journey from Hampstead Heath to the Thames. London has twenty other lost rivers, including the Fleet, the Black Ditch, and the Tyburn.[4]

After bringing a river back into the light, the next step is to clean up pollution. Some small London tributaries are coming back to life, including the river Wandle, which passes just a mile from where I live in South London.[5] But cleaning up the Thames remains a challenge. New sewage-treatment works brought salmon and other fish back to the tidal Thames half a century ago, but a growing city population and slow investment have seen those improvements reversed. "We have been to hell and back on the Thames," says Alastair Driver, of the Environment Agency. "For a few days a year, it becomes disastrous. You can see sewage in the river for 10 kilometers upstream of London Bridge." In 2017, Thames Water, the company that operates the sewage works, was fined £20 million for routinely pumping thousands of acre-feet of raw sewage into the Thames from overloaded treatment works in Oxfordshire.[6]

After cleaning pollution, bringing back natural flows is next on the restoration checklist. Europe's water demands are not so great that rivers generally run dry. But disruption to their natural hydrology is intense. Dams, weirs, and other barriers everywhere disrupt fish migrations and change river flows. Hans Bruyninckx, director of the European Environment Agency, says his scientists have counted half a million human-constructed barriers across rivers in Europe. That is almost one every mile.

Besides tearing down barriers, river restoration requires re-creating old channels and meanders, revegetating banks, and reconnecting rivers with their floodplains. The task is huge. Northern Sweden may appear unpopulated and largely untouched by humans. But from the 1850s to the 1970s, foresters straightened and cleared vegetation on huge numbers of rivers so that logs could be floated downstream to ports. Now those rivers are being restored. Oddly, the main technique is dumping trees in the river. "We need large structures—they are very important for slowing down water and trapping organic material," Johanna Gardeström, an ecologist at Umea University, told me. "The trees are also food for aquatic insects, which in turn are food for fish."

Trees are also important because their shade keeps rivers cool. "Salmon and brown trout die if water temperatures stay above 72 degrees Fahrenheit for over seven days," says Rachel Lenane of England's Environment Agency. "With climate change, we are seeing those conditions in southern England now." But trees can cut 9 degrees off water temperatures if there is full cover, she says.

Some of the most dramatic environmental battles in Europe have been over water engineering projects, most notably on the Danube. Back in the 1980s, the Soviet-inspired Gabcikovo-Nagymaros Project—aimed at improving navigation, preventing flooding, and generating hydropower for Czechoslovakia and Hungary—helped bring down the Berlin Wall. Massive opposition in Hungary to plans to build the Nagymaros Dam on a much loved bend in the Danube helped cause an upwelling of more general political opposition that eventually brought down the Communist government in Budapest in 1989. It was the first brick out of the wall.[7]

The Danube, which runs west to east, from Germany's Black Forest to its delta on the Black Sea, is the most international river in the world, with a catchment that includes nineteen countries. The river has been cut off from 80 percent of its floodplain. Today much of that floodplain is slated for restoration, however. Slovakia, the Czech Republic, and Austria are together putting meanders back into a canalized Danube tributary, the Morava River, where NATO forces on one bank once eyed Warsaw Pact forces on the other. Austria and Germany have been removing levees to restore the floodplain of another tributary, the Inn River, at the foot of the Alps. And downstream, Ukraine has taken down levees on two of the largest islands of the Danube Delta, allowing spring flooding, the return of birdlife, and the introduction of free-roaming cattle.

"Until recently, the model for our engineers was to have straight rivers," says Gheorghe Constantin, director of water resources in Romania, another Danube nation. "We kept building levees right up to 2006, when there were huge floods on the Danube. The levees broke. So we decided to leave more space for the river. It was a new model, taken from the Netherlands."

One of the last free-flowing stretches of the upper Danube, between Vienna and Bratislava, the capital of Slovakia, is managed by a new Austrian national park. There, wardens are restoring lost side channels, as well as bankside forests. "We want to end the fortification of the Danube," said Carl Manzano of the Donau-Auen park authority. "We are taking away concrete and riprap so the river can re-create its natural bank. About 50,000 cubic meters of stone structures have come down." Kingfishers are returning, and so are wild bees and birds like the little ringed plover. But, he admits, full restoration requires much more. Naturally, the river would break up into many channels, he says. But "we have to accept" that a full restoration of the natural river may never be possible.

Europe's river restorers face practical limits on what they can achieve. Ulrich Pulg of Uni Research in Bergen, Norway, estimates that half of the rivers in Germany, a third of those in Norway, and two-thirds in Belgium can never have their floodplain ecosystems restored, because whole cities would have to move. While the restorers work out what they can achieve in some areas, they say they are still losing ground in others. Ill-advised engineering projects continue. To the restorers' frustration, some are being carried out in the name of the environment.

For example, to fight climate change, hydroelectric-power generation is making a comeback in Europe. It is seen by some as the ideal method of generating low-carbon energy. Sweden already gets more than 40 percent of its power that way, Austria 60 percent, and Norway more than 90 percent. Countries in the Balkans now want to go the same way and are keen to make up for lost time. In 2016, there were more than six hundred dams proposed in what Ulrich Eichelmann of the campaign group Riverwatch calls the "blue heart of Europe," between Slovenia and Albania, where near-pristine rivers survive.

An early victim of this frenzy of concrete pouring could be the river Vjosa in Albania. In 2016, more than two hundred scientists from thirty-three countries called for the Albanian government to halt plans to construct the first dam on the river, which flows for 400 miles out of northern Greece and through remote Albanian mountain canyons to the Adriatic Sea. Its ecology is both pristine and largely unexplored. "We might know more about rivers in the Amazon basin than about the Vjosa," says Eichelmann. The proposed

82-feet-high dam would be hugely damaging, particularly for birds that breed on the islands downstream of the proposed dam site at Poçem.

In mid-2017, judges held up the project because of a deficient environmental-impact assessment.[8] But the government has plans for another seven hydroelectric dams along the Vjosa. "A developing country cannot be a museum," said Albania's energy minister Damian Gjiknuri. "Hydropower has drawbacks, but every development has a cost to the environment." Eichelmann wants instead to create "the first European Wild River National Park."

Other East European countries continue to pour concrete and chop down bankside trees. They are not building large dams; they are rebuilding their rivers in the belief that this is the way to improve flood defenses. I went to see this in Poland, where the conservation group WWF revealed in 2013 how a potent mix of European Union funding and old Communist engineering habits were creating an environmental disaster.

Across Poland in the previous five years, some 10,000 miles of rural streams, a fifth of all the country's rivers, had been "improved" by engineers: culverted, dredged, straightened, fitted with concrete banks, cleared of vegetation, and diverted. It looked like a blatant contravention of the European Framework directive's call to return rivers to a "good status." Yet it was being financed with tens of millions of euros from Brussels that had been earmarked for developing the country's infrastructure. How to kill a nation's rivers? This was how.

Southwest of Warsaw, Marcin Kazmierczak, an environmental engineer by profession and conservation activist in his spare time, took me to the river Moszczenica. In a valley lined with trees, the natural course of the river had been replaced with a straight concrete-lined canal. "They simply landfilled the old river," Kazmierczak said. "This was my favorite spot, the most beautiful part of the river. Before this engineering, it had beavers, otters, and weasels. But they are gone now." The canal was intended to protect local farms from floods, he said, though in the steep valley, it was hard to see who would benefit.

We drove on to the river Drzewiczka by the village of Ogonowice. The natural river had been dredged and the bank cleared of vegetation and covered in rocks. Tree roots that stabilized the bank had been cut down to make way for the stone riprap. Was this protecting the village from floods? Przemek Nawrocki, a veteran campaigner with WWF in Poland, thought not. He had published a poster that read, with savage irony, POLAND WILL SOON BE THE PERFECT SITE FOR RIVER RESTORATION . . . BECAUSE MOST OF ITS RIVERS WILL BE DESTROYED.

According to the European Environment Agency, the best way to solve Europe's growing problem with river floods lies not in turning its rivers into concreted floodways but in reviving their natural floodplains.[9] There have been more than thirty-five hundred floods since 1980, the agency says, and the trend is upwards. In 2010, no fewer than twenty-seven countries were affected. One cause is more intense rainfall due to climate change. But much of the actual damage arises because rivers have been cut off from their natural floodplains.

On many major European rivers, more than 80 percent of the low-lying land, where rivers would naturally spread during high flows, has been barricaded off by flood defenses, then drained and built on. In England, 90 percent of floodplains "no longer function properly," a 2017 study found.[10] In Hungary, almost a fifth of the population lives on what was once floodplain. The barriers may protect some stretches from floods, but they raise river levels higher everywhere else. Somewhere, the river will inevitably burst its banks, flooding homes and streets. The flood will be fiercer than it would otherwise have been. Without action to curb counterproductive flood defenses, the EEA report predicted a fivefold increase in flood damage by 2050.

For too long, engineers saw rivers as little more than navigation routes, and pipes to supply water, remove waste, and rush floodwaters to the ocean. Nature was an inconvenience that had to be tamed. But if you pick a fight with nature, you usually lose. Take the case of the Rhine. Attempts to tame the river began in earnest in the nineteenth century, with "rectification" works undertaken by German engineer Johann Tulla. The very word "rectification" indicates the extent to which engineers' notions about re-creating the world on their terms had taken hold. Nature had got it wrong; we had to "rectify" things.

Until then, the upper reaches of the Rhine took a circuitous, wandering path across a wide floodplain of woods and water meadows. Between Basel in Switzerland and the German industrial city of Karlsruhe, the river split into innumerable branches that continually moved, disappeared, and reformed. The islands between the branches were covered by flooded forests and wet cattle pastures. During each spring flood, the silty Rhine water slipped into the forests, meadows, and fields.

It sounds idyllic, but the split channels and snaking route prevented the passage of all but the smallest boats, and building was difficult on shifting riverbanks. Even more disconcerting, the border with France, which followed the main channel of the river, moved whenever a flood passed. So Tulla forced the river into a single, well-defined, permanent channel. "As a rule," he declared, "no stream or river needs more than one bed." Nature never intended that this should be so, but Tulla's maxim has become the rule that almost every river engineer follows.

The rectification prepared the Rhine to become the great river highway of Germany. Along its banks grew industrial cities such as Mannheim, Koblenz, Cologne, and Düsseldorf. More than two thousand river islands in southern Germany disappeared, and most of the sluggish backwaters and shallow gravel reaches disappeared. The new, straight upper Rhine was about 60 miles shorter than the old river, and it flowed a third faster.[11]

But this was only a halfway house to the modern Rhine. Tulla's plan kept high dikes well back from the river, maintaining much of the floodplain. As time passed, however, farms, villages, and towns added their own dikes nearer the stabilized river, with the aim of protecting buildings and turning seasonally flooded pastures into arable fields. During the twentieth century, an entirely engineered channel, complete with giant locks and a series of hydroelectric plants, finally cut off the upper Rhine from 60 square miles of its floodplain, leaving just a tenth of the original area available over which the river could flood.

The faster river sped barge traffic to the sea but also halved to just thirty hours the time it took floods to pass out of the Alps and travel downstream from Basel to Karlsruhe. After heavy rains, the peak flow on the Rhine now coincided with peak flows in tributaries such as the Neckar as they met the main river. The resulting flood surge rushed downstream toward Bonn and Cologne, where peak flows were a third higher than before. Floods that previously were likely once every two hundred years could now to be expected every sixty years.

On top of that, the new, faster river scoured the river's bed and banks much more fiercely. This had been exacerbated because the artificial embankments on the remade Rhine prevented the river from collecting silt from its former upstream floodplains, so it was much more able to erode material from the riverbed as it traveled downstream. At Basel, the bed of Tulla's rectified river had fallen by 23 feet; at Duisburg it had fallen by 13 feet.

Tulla was known in his time as the "tamer of the wild Rhine," but his efforts helped turn the old, slow, silty stream that laid down alluvium as it went into a fast, silt-starved, and eroding river. Almost two centuries of efforts to tame the Rhine had made it wilder and more unruly. And far less able to handle extreme floods when they come roaring out of the Alps.

The results were inevitable. During floods on the Rhine in 1995, levees failed and large parts of the Netherlands at the river's mouth flooded. It was a disaster that hogged the world's headlines, and an existential crisis for a Dutch nation largely built by reclaiming land from the sea. A quarter of a million people were evacuated in the Netherlands alone. In the aftermath of the floods, the country boldly decided that confronting rivers did not work. The rivers had been straightened and the land had been built on so much that there was nowhere for the water to go. However high you raise the levees, a river in flood will find the weakest spot and burst through. Rob Leuven, of the University of Nijmegen, said at the time, "Government policy is now to allow rivers to take more physical space." Rivers were to be allowed back onto their floodplains, and overspill areas of low-lying land would be set aside to lessen the impact of major floods.

The Dutch are caricatured as always having their fingers in their dikes, but now they have given engineers the go-ahead to punch holes in the dikes. They are making plans to return up to a sixth of the country to soggy nature in order to better protect the rest. By 2015, they had finished thirty-four such projects at a cost of more than $2 billion. The Germans have also given top priority to finding ways of lowering the floods on the Rhine. In 2002, environment minister Jürgen Trittin introduced legislation that would "give our rivers more room again; otherwise they will take it themselves."[12] The target is to shave more than 23 inches off the flood peaks by 2020, by reinstating some of the 500 square miles of floodplain lost on the lower Rhine. Drained and diked fields will be replaced by reed beds and water meadows designed to flood every winter.

37

GANGES AND MISSISSIPPI

Floods from the Sea

THE RAMSHACKLE RIVER PORT of Khulna in southwestern Bangladesh is one of the most flood-prone urban areas on earth. The third-largest city in one of the world's poorest and most populous nations is at constant risk of inundation. It lies 80 miles inland from the shores of the Indian Ocean on one of the giant natural waterways that discharge the waters of the river Ganges into the ocean. The land is low-lying and surrounded by ponds, where farmers raise prawns. The city makes millions of dollars a year for the industry's criminal gangs, known locally as the "prawn mafia." But despite flood-protection banks, a tenth of this city of two million people is permanently waterlogged, and floods occur on average ten times a year.[1]

The conventional explanation for this quagmire is that rising sea levels due to climate change are leaving such places at ever-greater risk from high tides. That is only part of the story, however. Sea levels in the open ocean are indeed rising—by more than an inch a decade. But high tides in Khulna and along the other river estuaries of southwestern Bangladesh are rising five times faster. There is a surprising reason for this, say hydrologists: misguided flood protection.

Traveling by boat on the great river estuaries of southern Bangladesh is an overwhelming experience. Some of these waterways are wider than the English Channel. Between them are wide and constantly shifting delta flats at permanent risk of inundation. Floods that would be front-page news in other countries barely register here. During my visit, thirty villages disappeared under water, after a 12-mile length of embankment collapsed. That made three paragraphs in the national media. "Several Thousand Marooned for the Seventh Time in the Last Four Months," ran another routine headline.

Once many of the islands and flatlands were protected by mangrove swamps that prevented erosion and absorbed high tides. But in the past century, successive governments have allowed hard-pressed farmers to chop down these natural flood defenses. Governments have sought to replace them with around 2,500 miles of earth embankments enclosing low-lying land known as polders. During the 1960s and 1970s, USAID paid for a thousand miles of embankments in the Khulna district alone. Those banks surround more than a million acres in forty polders, where about thirty million people now live.

There is mounting evidence that the embankments that seal off these polders from the wider delta may make things worse. They are constricting incoming tides. This funneling effect amplifies the tidal range and pushes tides farther inland, making them much higher than before in places such as Khulna. As a result, "extremes in tidal water levels are more prominent in inland areas compared to those near the sea," Shahjahan Mondal of the Bangladesh University of Engineering and Technology reported in 2013. In southwestern Bangladesh, the "maximum tidal high water level is increasing at a rate of 7–18 mm a year," which is between twice and six times the increase in sea levels in the open ocean.[2] The highest tidal range was recorded at Khulna.

In a more detailed study for the World Bank, British geomorphologist John Pethick found the same on the Pussur Estuary near Khulna. High tides had been rising 6 inches a decade.[3] He too blamed the embankments and called the flooding "mainly a self-inflicted wound." He predicted that the combined effect of the funneling and the global rise in sea levels could result in a staggering 8-foot rise in high tides in the delta by the end of the century. The more the government raised banks to try and protect against floods, the worse the problem would get.

That sounds just too bad to be true. But I found plenty of support for Pethick. "The data show an increase in the tidal range, and the impact of poldering is a reasonable explanation," Michael Steckler, of Columbia University in New York, told me. Tidal amplification is common in estuaries and river deltas around the world, agreed coastal engineer Marcel Stive of Delft University of Technology in the Netherlands.

Pethick says Bangladesh has no alternative but to tear down some of the flood banks to give the rivers more room and to set the others farther back, to reduce the funneling effect. He would re-create the old natural defenses, such as the mangrove swamps. Such suggestions have angered many Bangladeshis. This is understandable. Flood engineers don't like to be told that their work is making things worse. The country's politicians see Western scientists making excuses for the rising sea levels caused by Western pollution. They don't want to give up land to rising tides caused by others.

So, even though much of Pethick's work was carried out as a consultant for the World Bank, the Bangladesh government and the World Bank are going ahead with a $400 million upgrade of some 400 miles of embankments south of Khulna. The project is due for completion in 2020.[4] Pethick is frankly incredulous. The current policies, he says, are not protecting people from floods. They are putting millions more people at risk.

~~~

For a sign of what Khulna may face, we need look no further than New Orleans. In 2005, it too faced a lethal combination of a tidal surge from the ocean combined with disastrous attempts to curb floods on a major river's low-lying delta. In this case, the river in question was the Mississippi, which drains the planet's second-largest river catchment.

When the Spaniards first arrived on the river's floodplain in the sixteenth century, they thought better of building on it. They left great expanses of marshes and lakes to the Native Americans, who had long experience of coexisting with the floods. But the French settlers who followed felt no such constraints. They founded New Orleans right on the river delta and called their marsh colony Louisiana, after their king.

New Orleans was always trouble. Water filled the basements of its buildings in the first months after its founding. Over the years, the French raised the natural levees ever higher. After an inundation in 1735, they built earthworks for 45 miles around the city. Then when the United States bought the lower river as part of the Louisiana Purchase, slaves from dozens of cotton plantations, equipped with shovels and wheelbarrows, raised hundreds of miles of banks to speed "ol' man river" to the sea. But still it flooded.[5]

In the mid-nineteenth century, Congress passed the Swamp Land Acts, which handed ownership of unclaimed swamp to state governments. They were allowed to sell the land to raise money for further levee building. This brought to the Mississippi a land-drainage fever that had already taken hold in Europe. Louisiana sold off a quarter of the state to farmers, who drained the swamps for plantations. In so doing, however, they undermined the effect of the levee construction that the sales were funding. For, in hydrological terms, the swamps were giant holding reservoirs for the river's natural floods. They spread the river's flow through the year, holding on to floodwaters during high flow and letting it seep back during low flow. Now barricaded off, they could do neither. The river floods grew higher, and the levees broke with increasing regularity.

Congress upped the stakes again in 1879, handing control of the river it-self to the United States Army Corps of Engineers through the Mississippi River Commission. The aim was to wage war on the river and finally "prevent destructive floods." But the river gave its reply by breaking the levees in 180 places in 1882. In 1927, despite ever greater expenditure by the corps, Missis-sippi floods submerged an area the size of Belgium, destroying 160,000 homes and every bridge for 1,000 miles upstream of Cairo, a small Illinois town with a name as presumptuous as the engineers' science. That year, New Orleans was saved only when, recognizing the river's unstoppable force, corps engineers blew up a levee to draw off the flood.

Since 1927, some $7 billion more has been spent on raising levees and southern Louisiana has lost a fifth of its wetlands, some 2,000 square miles.[6] The trouble is that separating the river ever more firmly from its floodplain has only increased the flows during major floods. The 1993 floods demonstrated that. Much of St. Louis was submerged beneath water that rose 50 feet above the normal level. The river rose over more than 600 miles of levees. Nearly five hundred counties featured in a presidential disaster declaration, and property damage was put at $12 billion.

Then came Katrina. The hurricane that battered New Orleans and the Gulf Coast at the end of August 2005 came from the sea. But its most lasting im-pact arose from storm surges that penetrated deep into the Mississippi delta. Because the Louisiana wetlands had mostly been drained, and levees raised to prevent water in the river escaping across the delta, the water surging inland from the Gulf had nowhere to go. Water levels rose and rose until finally they breached New Orleans' levees, engulfing a city of half a million people.[7]

One of the river's nineteenth-century engineers, James Eads, uttered one of the great clarion cries of river engineering when he declared, "Every atom that moves onward in the river, from the moment it leaves its home among the crystal springs or mountain snows, throughout the 1500 leagues of its devious pathway, is controlled by laws as fixed and certain as those which direct the ma-jestic march of the heavenly spheres. . . . The engineer needs only to be insured that he does not ignore the existence of any of these laws, to feel positively certain of the results he aims at."[8] In practice, the river has proved consider-ably more intractable. Mark Twain was perhaps closer to the truth when he observed, after the creation of the Mississippi River Commission to hold back floods, that "ten thousand River Commissions, with the mines of the world at their back, cannot tame that lawless stream, cannot say to it, 'Go here,' or 'Go there,' and make it obey."[9]

38

MORE CROP PER DROP

GEOFF HEWITT had a big new house and a four-car garage. His spread on the Dalby Downs, a four-hour drive inland from Brisbane, the capital of Queensland, Australia, had 5,000 acres of cotton and enough trees down by the creek for the roosting kookaburras to wake him in the morning. He had high-tech, GPS-navigated tractors, able to steer to an accuracy of an inch. He had reservoirs able to hold 5,000 acre-feet of water to irrigate his fields. But something was missing. Crops. Outside on the veranda, listening to the kookaburras laugh, Geoff explained why, months into the growing season, not a single acre of his farm was planted. "We've had three and a half inches of rain so far this year. Usually we get twenty-five inches in a year, and it's September." The creek was dry; his reservoirs were empty. So there was no point in planting cotton.

Hewitt was at the heart of a crisis that had engulfed the Murray-Darling river basin, Australia's largest and stretching for a thousand miles from the Dalby Downs to Adelaide in South Australia. Usually it was where half the country's crops grew. As we looked out across his parched fields, he pointed toward the dried-up creek. "That is the start of the Condamine River," he said. "It is one of the sources of the Murray-Darling. Or should be."

Yet as I probed further, I got the impression that even in this long drought, Hewitt was oddly wasteful of water. He had some of the most sophisticated, computerized methods available for delivering just the right amount of nutrients to the roots of crops. But he didn't use the same sophistication in irrigating his crops. When he had water, he simply flooded his fields and presumed some of the water would reach plant roots. Even more surprising, his reservoirs were so shallow that, when I asked him to do the math, it turned out he was losing up to a third of his water to evaporation. He was surprised; he'd never thought to do the math before.

He was not alone in his water-blindness. Australia's fifteen hundred cotton farms were using, when they could get hold of it, more water on their farms than the country's seven million householders. The Cubbie Station, 25 miles by 12 miles, west of Brisbane was the largest irrigated farm in the Southern Hemisphere. Its reservoirs had the capacity to store more water than Sydney Harbor. But they only seemed to care when the water runs out.

This was 2006. The basin was in the grip of a drought that had already lasted more than a decade. A combination of little rain and the ruthless abstraction by farmers had left gum trees all down the river dying in their tens of thousands, and with them the ecosystems they sustained, everything from cormorants and pelicans to rare squirrel gliders and carpet pythons. The Murray's mouth had become choked with sand as the last flow came to a halt in a small, increasingly saline pool within the once huge lagoon of Lake Alexandrina. Only dredging maintained the lake's link with the sea.

The following year, after a decade of badgering by the country's ecologists, the government concluded that enough was enough. Prime Minister John Howard, an arch conservative, announced a plan to rein in water demands and give some water back to the river. "No other single substance has a greater impact on the human experience or on our environment. . . . We need to make every drop count," he said. At the time of writing a decade on, the Murray-Darling Basin Plan he launched that day is still a work in progress. But most of the target of returning 2.2 million acre-feet of water annually to the river has been achieved.[1]

Districts along the river and each farm have a water allocation given each year as a fixed proportion of the water flowing down the river in that year. In wet years they get more water, and in dry years less. The environment also gets a share. The rights can be traded, and with a finite amount of water available, most users strive for efficiency. Geoff and his fellows on Dalby Down now have an incentive to save water.[2] In late 2016, the river saw its highest flows through South Australia since 1993. Salt that had been accumulating on the floodplain for two decades was flushed out. The billabongs and floodplains were wet again, nature burst back into life, and farming communities breathed again.[3]

~~~

Since the first edition of this book was published in 2006, the drive to conserve water and to make more effective and economic use of water has grown. What Australia's farmers have done on the Murray-Darling basin, while incomplete, has become a marker for what is possible when governments take firm action. But there is a long way to go.

Farmers across the world have traditionally irrigated their fields by pouring on water indiscriminately. Some of that water reaches crop roots; most does not. It either evaporates in the sun or seeps underground. The result has often been waterlogged soils, a buildup of salt, and growing water shortages. At a conservative estimate, two-thirds of the water sent down irrigation canals never reaches the plants it is intended for. Technologists early in the twentieth century got to work on the problem. First they came up with the sprinkler, which doused fields in a fine spray of water from a central pivot. That saved the worst excesses of overirrigation but was equally indiscriminate and lost a lot of water to evaporation. It was also expensive and required energy to spray the water.

The next advance was drip irrigation. Credit for this idea is normally given to an Israeli engineer, Simcha Blass, who retired to the Negev Desert in the early 1960s. As he told it, one day he noticed how a large tree grew in the desert because it was right next to a slowly dripping faucet. He mused on the matter and subsequently invented a narrow tube that could deliver water under pressure and drip it close to the roots of plants. He filed a patent in Tel Aviv in 1969. His idea coincided with the development of plastics that made such a field-water distribution system economical and with the growing realization in arid countries such as Israel that water was in increasingly short supply.

Drip irrigation can take many forms. Mostly it is high-tech, with water pumped down pipes under pressure and sent into side pipes from which sophisticated "drippers" deliver it to roots. Such systems may include flow meters, pressure gauges, and even soil-moisture measurement to optimize delivery and keep losses to a minimum. Today large farms in water-stressed regions such as California, Tunisia, Israel, and Jordan specialize in such systems. In Jordan, drip irrigation has reduced water use on farms by a third while raising yields. Israeli farmers have increased water productivity fivefold, through a mixture of drip irrigation and the recycling of treated urban wastewater onto their fields.

The technology should have taken off worldwide. But it remains virtually ignored by the mass of small farmers in poor countries, who face some of the worst water shortages. In India, despite strong government promotion and subsidies, drips irrigate only a tiny minority of fields. This is perhaps not surprising, since the full kit can cost more than $800 for one acre, and even stripped-down systems developed for poor countries cost at least $200. Another reason is that most farmers in most places at most times get their water at heavily subsidized prices—a tenth of the real cost is typical everywhere from India to Mexico and Pakistan to California. When farmers pump water from beneath their fields, they pay no more than the cost of a pump and electricity, which itself is usually highly subsidized. There is little incentive to save water. So what will turn the tide?[4]

Top-down state control is one way. As its rivers ran dry and fields turned to dust, China in 2009 announced dramatic plans to cut water use by industry and agriculture. Water resources minister Chen Lei said the country would reduce the amount of water needed to produce each dollar of GDP by 60 percent by 2020. With the economy on course to grow by 60 percent by then, that effectively meant he wanted to plateau water consumption with immediate effect.

China irrigates more land than any other country. Around half of its farmland relies on it. China's farms use about 55 percent of the country's water supplies. However, the proportion is falling as cities demand more. The government's assessment of water efficiency on farms has improved from 44 percent in 2004 to better than 50 percent by 2015, with a target of 60 percent by 2030. To achieve that, the proportion of farms using water-saving irrigation technologies of one sort or another will rise from half to three-quarters.[5]

Another route to water efficiency may be bottom-up, farmer-led systems, following on the rising trend for farmers to manage their own water resources rather than relying on state irrigation systems. Some think Pepsees can save the world. Pepsees are small tubes made of light, disposable plastic. They were originally designed and sold to encase individual ice candies—freeze-pops, or whatever you choose to call them. They are manufactured all over the world and are just about the most disposable product imaginable. Millions of roadside vendors across the land buy the polyethylene tubes, which come in long rolls that can be torn off at perforations stamped into the plastic every 8 inches or so. In India they are made under the Pepsee brand name.[6]

Sometime around 1998, somewhere in the Maikal Hills of central India, someone—perhaps a farmer with a sideline of selling ices—started using Pepsee rolls for another purpose: to irrigate the fields. The farmer had discovered that the rolls of plastic tubing made perfect cheap conduits for distributing water to plants. Shilp Verma, of the International Water Management Institute in India, says, "It is not very clear how and exactly where the innovation first started, but it spread among farmers like wildfire."[7] The farmers took the rolls, laid them down rows of crop plants as close as possible to the roots, and poured water in at one end. As the water ran down the tubes, it dripped through the perforations to water the plants. This startlingly successful exercise in lateral thinking has produced the first dirt-cheap method of providing drip irrigation for poor farmers. Twenty years on from its invention, its progress remains limited, but it shows what can be done.

Technical innovation is vital. There is plenty to be done at plant-breeding stations. Most modern high-yielding plant varieties are hopeless water guzzlers. Only now are breeders turning their attention to producing varieties that use water efficiently. Rice, the world's most popular grain, consumes more water than any other food crop, typically twice as much as wheat for every ton of output. More than a third of all the water abstracted from rivers and groundwaters on the planet goes to irrigate rice paddies in Asia. But with half of the world's rice grown in water-stressed China and India, that level of production cannot continue.[8]

New rice varieties could one day cut water use by as much as half. As well as new crop varieties, new methods of growing the crop could help. Experiments in India and at the International Rice Research Institute in the Philippines show that rice can be grown with much less water. The trick is to abandon the traditional method of growing seedlings in nurseries and then transplanting them into paddies that have to be kept flooded. Much less water is needed if they are instead planted directly into muddy soil. Some varieties can already be grown this way. If those varieties were more widely adopted, they would cut water use by a fifth—and require less labor in the bargain.

The lesson here, I think, is that we are currently almost unbelievably inefficient in the way we use water. From the millions of rice paddies across Asia to Hewitt's high-tech cotton operation, everyone has taken water for granted. But if we are so bad at managing water, then there is still massive potential to do better. Just as half a century ago the world managed the seemingly impossible by doubling crop yields per acre of land, the need now is to double yields per acre-foot of water. Because water rather than land is now the more precious resource. After the green revolution, we need a blue revolution.

We need to rethink water not just on the farm and in the laboratory but in the world of river management and infrastructure construction. The dam builders, too, live in a world of finite resources.

More and more rivers are becoming fully utilized, sometimes several times over. In hydrologists' jargon, these rivers are known as closed basins. The only water left for nature is often drainage from humanity's last use. Every effort to utilize the water better—whether building dams or lining canals or improving irrigation efficiency or even harvesting the rain—is not providing more water but simply taking from one user and giving it to another.

Water engineers, agriculturalists, and even environmentalists continue to promote the idea that if farmers "save" so much water by pouring less of it onto their fields, then the entire valley or river basin will have "saved" that figure. If they prevent evaporation, that can be a genuine gain. But the same is not true for water that is "saved" by preventing it from seeping underground. Because when that happens there will be less water for others to pump up, or to return to a river for downstream use and to sustain its fluvial ecosystems.

We saw this on the Rio Conchos in northern Mexico, where lining irrigation channels just reduced the amount of water reaching the aquifers. Similarly, Spain spent heavily on improving the efficiency of its irrigation system during the 2000s, in the hope of staving off the environmental impacts of declining rainfall. It discovered, however, that water "returns" to rivers from seepage were much reduced, so there was less water to pump out from rivers farther downstream.[9] The same unintended outcome has been documented in western Kansas, where efforts to save the Ogallala Aquifer with state-subsidized conversion to more "efficient" sprinkler irrigation resulted instead in increased groundwater loss.[10] One man's waste of water is often another man's source of water. In many places, things end up worse than before, because farmers who had improved the efficiency of their operations were allowed to irrigate more fields, says Bruce Lankford of the University of East Anglia.[11] It seems obvious, yet across the world the simple insight remains remarkably elusive. Water is the ultimate renewable resource. But that only makes the task of managing the water cycle prudently more critical.

39

COLOMBIA

Dollars and Sense

CLIMBING A STEEP, ancient track from the tiny, whitewashed colonial town of Cácota high in the Colombian Andes, I could feel the air becoming thinner. Soon I was in fog—slipping on the wet stones as I left the cloud forest and headed higher into the mountains. Farmers near the track were cultivating potatoes on improbably steep hillsides. But mostly our party had left humanity behind as we finally reached a region of alpine plants known as the páramos. With swirling winds and constantly moving fog, this treeless, sunless world is an improbable ecological El Dorado stretching from 9,200 feet in elevation to the snowline at over 13,000 feet. It is Colombia's wet hydrological sponge, the source of most of its rivers, its water supplies in Bogotá.

If it weren't for the lake, just visible through the fog in front of me, I could have mistaken this landscape for desert. It was covered in weirdly shaped, cactus-like plants—most of them found nowhere else. They are xerophytic plants, meaning they are adapted to desert conditions. Yet water is the one thing this landscape did not lack. Bathed in near-constant swirling clouds, the plants were sopping wet and their roots sat in saturated soils. "The whole place is a natural reservoir," said my guide Sergio Ivan Niño of the local conservation authority, as a gentle drizzle began to fall. The water was captured in the saturated soils and seeped slowly into lakes, peat bogs, springs, aquifers, and eventually rivers.

The Andean páramos have no parallel on this scale anywhere else in the world, says Wouter Buytaert of Imperial College London, who has studied them more than most. They cover an estimated 17 million acres in Colombia, Ecuador, Peru, Venezuela, Bolivia, and Costa Rica. The only others are "a couple of patches on Kilimanjaro in East Africa and on the mountains of New Guinea." They are the major reason why—as every visitor is told on the in-flight video before arrival—Colombia ranks seventh in the world for to-tal water resources and fifth for biodiversity. The spongy soils, which literally bounced as I walked across them, contain more moisture than almost any other soils, anywhere—typically up to 30 pounds of water in a cubic foot of soil. They cover just 2 percent of Colombia's land area yet provide 90 percent of its water and generate 60 percent of its electricity at hydro dams. "They are the key to the hydrology of the country," said Niño.

But the páramos sponges are feeling the squeeze, thanks to climate change and land invasions. According to the Colombian government, 75 percent of them could disappear this century.[1] Already, a decline in cloudiness is causing evaporation rates to rise high in the Andes, and rainfall is becoming more sea-sonal. Meanwhile, farmers are moving farther up the mountainsides, annexing parts of the sponge.

On the road to Cácota, I passed trucks laden with potatoes and milk churns left beside the road for dairies to collect. On my walk up the ancient track into the páramos, I saw plowing on 30 percent gradients. This is agriculturally unsustainable and ecologically insane. If it continues, the earth will erode and millions of tons of sediment will flow into the rivers that supply drinking water to the cities below. As the earth is lost the water-storing capacity of the páramos will disappear. The farmers need some incentives to change their practices. And that, improbably enough, is where beer enters the picture.

Leaving the plane at Bogotá airport, the first signs you see beside the mov-ing walkway to the baggage-claim area are advertisements for Aguila beer. It is the top brand from Colombia's monopoly brewery, owned by the people who supply America with Miller Lite. With the local market sewn up, the compa-ny's big concern today is not beer but water. Brewers have always been touchy about water. It takes a lot of H_2O to make beer, and that looks bad if it comes at the expense of secure supplies of nonalcoholic domestic water supplies. The makers of Aguila would rather be seen as helping conserve water supplies than emptying them.

So in 2009, the brewery partnered with the Nature Conservancy to help establish the Bogotá Water Fund, with the aim of raising $60 million from prominent water users to invest in conservation activities in the páramos.[2] The fund pays landowners to refrain from farming and offers grants to cattle raisers to replace their old herds with higher-quality cattle that produce more milk on less land. That, they hope, will protect the forests, soils, and páramos; keep sediment out of the rivers; and sustain water flows for more than eight million people downstream. And a brewery.

Colombia is a perfect laboratory for this idea of corporate finance for water conservation. For over half a century, the nation has been in a state of civil war, with the FARC guerrilla group and drug barons taking over much of the country—including the páramos. The violence kept development away. But now that peace has broken out, much could be lost up in the mountains.

How much difference can it make? That is hard to tell. The fund was handing out cash, but it is hard to see any positive return in spared land. Moreover, agriculture is far from the only threat to the páramos. There is a lot of gold and silver in the mountains, and mining is arguably a much bigger threat to the alpine sponge than farmers. The government also declared support for maximum exploitation of the country's mineral resources. In theory, Colombian law prevents mining in the páramos. The government's definition of the páramos seems flexible, however. The mining ministry has given the go-ahead to mining in areas that the Humboldt Institute, a research offshoot of the Colombian environment department, has decreed fall within the páramos zone.

The biggest of these is the La Colosa project, reportedly the seventh-largest undeveloped gold reserve in the world, promoted by London-listed AngloGold Ashanti. The government has issued exploration permits covering 240 square miles. In 2017 the mines minister insisted the mine should go ahead, despite a referendum showing intense local opposition. He denied that rivers were at risk. But Carlos Sarmiento of the Humboldt Institute strongly disagreed. The scheme would wreck large areas of páramos, destroying their water-holding capacity, he said.[3]

Maybe the páramos are doomed after all. But optimists say that the new system of payments has recruited the peasants to the cause of keeping the miners out. That certainly seemed to be the case in Cácota when I visited. At a meeting to welcome us, the villagers distributed peaches from their orchards and said they wanted to keep mining companies out, so they could invest the brewers' money in developing ecotourism. "We can only have prosperity if we protect the environment," said the mayor's spokeswoman. "Our natural riches are our village's strength." My walk up the ancient track could soon become a popular excursion for tourists, with the mysterious fog-shrouded lake as the prime destination. I'll drink to that.

Water's profile as an environmental issue has soared in the past decade, and corporate promises to go easy on the stuff have proliferated. Coca-Cola has been in the frame for water misuse since a row with villagers living near one of its bottling plants in southern India came close to wrecking its business across the country. In 2003, angry farmers in the village of Plachimada said the biggest bottling plant in Asia was drying up their coconut groves and paddy fields. The plant had for five years been taking around 130,000 gallons a day from boreholes on its 40-acre site. The story got taken up by the international media. The publicity was trashing the brand in the world's second-most-populous country. The truth is that even 130,000 gallons of water a day is not large on the scale of rural Indian water use. A single rice farmer with 25 acres could be using as much. Nonetheless, a Kerala court told Coca-Cola to stop pumping out the local water and the plant was shut two years later. The dispute was more about image than measuring water, and the company lost.

The salutary experience set the company, which uses 80 billion gallons of water a year worldwide, on a path to what it termed "water neutrality." Which in practice amounted to promising to "give back" through community water projects around the world all the water that it took for its bottling factories.

Other major food companies with large water footprints, like Nestle and General Mills, have signed up with environmental NGOs like WWF and the Nature Conservancy to an Alliance for Water Stewardship. Many are investing in water funds, especially in Latin America. Should we embrace these corporations as pump primers for conservation, or fear that we are seeing the start of the privatization of the hydrological commons?[4] After all, through their philanthropy, they are staking a kind of claim to ownership of these collectively owned resources. The Nature Conservancy is in no doubt that environmentalists should embrace the process. "Water funds are a part of TNC's major strategy to conserve large rivers round the world," it says. But they could end up alienating the custodians of the land and water—the local people essential to achieving conservation.

Take the first water fund in Africa. Launched in 2015, the Nairobi Water Fund aims to protect the upper reaches of the Tana River, Kenya's longest river. The river rises in the Aberdare Mountains and provides the Kenyan capital with some 95 percent of its water, while generating half of Kenya's electricity.[5] The new fund is masterminded by TNC and has money from Coca-Cola, the Kenya Electricity Generating Company, and the local subsidiary of another global drinks giant, Diageo, which brews Tusker beer in Nairobi. The fund's declared prime purpose is to educate upland farmers in ways to reduce the soil erosion that is silting up reservoirs and making the river's flow increasingly unreliable.[6] Fair enough. However, no representatives of farming communities sit on the fund's steering committee, which is dominated by big urban and industrial water users. It all looks a bit one-sided, with the small farmers as the fall guys.

The fund could easily become a convenient way for major water users to avoid answering questions about how they get access to the water they need while, as the fund's own publicity material agrees, 60 percent of Nairobi's residents do not have a reliable water supply. We live in a world where seemingly everything is open to being privatized and bent to the wheel of corporate profit. Should water really be added? Private profit could be a motive for improving the efficiency with which we use this most precious resource. But how far can or should this go?

40

WATER ETHICS

IT WAS RARE GOOD NEWS. On March 6, 2012, Ban Ki-moon, secretary-general of the United Nations, declared that the proportion of the world's population drinking unsafe water had been halved. One of the key Millennium Development Goals (MDGs) set by the UN more than a decade earlier had been reached five years ahead of schedule. Ban called it "a great achievement for the people of the world." There was just one problem. The claim was not true—and UN experts knew it.

A senior water professional with one of the UN agencies responsible for tracking progress to reach the drinking-water target told me that week, on condition of anonymity, "We should not say that the MDG water target has been met since we know that the indicator used to measure it has too many limitations." He called the UN claim "a drastic overestimate. We have not met the MDG target." Many other development and water experts within the UN system and among water NGOs agreed. They said the claim allowed UN officials to sign off with a successful outcome—but it hid far more than it revealed, including the failings of the UN and the international community.

The drinking-water target was set at the UN General Assembly in 1999, where countries promised to "halve, by 2015, the proportion of the population without sustainable access to safe drinking water." Halve, that is, from a baseline in 1990. For water professionals, the target created an immediate problem. Few developing countries measured the safety of drinking water. So the UN World Health Organization (WHO) and its children's agency UNICEF, which were jointly charged with monitoring progress on the target, decided instead on a proxy for access to safe water—"access to *improved* drinking water."

They defined "improved" water as coming from a piped supply, a drilled well, collected rainwater, or a hand-dug well or spring "protected" from sewage contamination. Unimproved sources included rivers and open wells and water delivered by carts, all of which face obvious risks of contamination. In effect, they replaced a health target with an engineering target. The UN said in 2012 that, by this measure, the proportion of the world's population without access to improved water sources fell from 24 percent in 1990 to 11 percent by the end of 2010.

But is it right to assume that an "improved" water source is safe? Sadly, the answer is no. A report on progress toward the target prepared by UNICEF and WHO a year before the same agencies declared success noted that the engineering proxy "does not guarantee the quality of drinking water consumed," and that lack of good historical data meant the agencies "cannot report on the actual water safety aspect of the MDG drinking water target."[1] The 2011 study found that more than half of water sampled from supposedly "protected" hand-dug wells was actually contaminated. So was about a third of the water from "protected" springs and drilled wells. As a result, in Nigeria and Ethiopia, two of Africa's most populous countries, only about two-thirds of the "improved" sources provided safe drinking water. In Ethiopia, only 27 percent of the population had safe drinking water, rather than the claimed 38 percent.

That should have been enough to call off all declarations of achieving the target. There were further concerns too. Not only was much of the water from "improved" water sources not safe, but many of the pipes and pumps supposedly delivering this water were broken. A 2010 study by UNICEF researcher Clarissa Brocklehurst estimated from UN and government data that 36 percent of hand pumps in twenty African countries, many of them installed to help the UN achieve its goal, were broken. She criticized the "shamefully" poor performance of drilled wells attached to hand pumps. A typical pump will break down within two years and most are abandoned within five. "Thousands of people who once benefited from a safe drinking water supply now walk past broken hand pumps or taps and on to their traditional dirty water point."[2] In the Menaka region of Mali, researchers found 80 percent of wells dysfunctional, and in northern Ghana, 50 percent.[3]

To meet the MDG target, governments and NGOs drilled wells, usually connected to simple hand pumps, but they paid little attention to keeping them functioning. Maintenance, said Brocklehurst, was regarded as "somebody else's problem." Many NGOs believe it is best to hand ownership of the wells to locals. But if the locals don't have the training or skills to do maintenance, or the funds needed to buy spare parts, then it is a recipe for failure. Even functioning pumps may be so far away that few people make the trek if there is another source, albeit a dirtier one, close by. Or there may be cultural taboos preventing their use. In India, for instance, in some cases only certain castes are allowed to drink from a particular well.

I wrote an article on all this in 2012.[4] It got a lot of flak, but the evidence kept mounting. In 2013, Mark Sobsey of the University of North Carolina at Chapel Hill found that 47 percent of improved water sources in the Dominican Republic weren't safe.[5] The following year, WHO's own journal published a paper by Joe Brown of the Georgia Institute of Technology in Atlanta, who said that often the only "improvement" delivered had been delivering the same dirty water in new pipes. Even in the official records of compliance, people receiving dirty water for a few hours a day, and none at all for the rest of the time, were counted as having water as safe as a householder in New York.[6] "It is quite unreasonable to assume that 'improved' equals 'safe,'" Brown told me. "The WHO has been silent about this."

WHO's head of water and sanitation, Bruce Gordon, responded that "we are well aware of the issues raised in the new paper." The organization hoped to do better monitoring in the future, using new low-cost kits that rapidly assess water quality. When I asked about the results of this in 2017, Gordon said that good progress was being made. But of course WHO could not retrospectively assess the safety of water. So we still do not know if the drinking-water target had truly been met.

The good news is that water is the ultimate renewable resource. We may pollute it, irrigate crops with it, and flush it down our toilets. We may even encourage it to evaporate by leaving it around in large reservoirs in the hot sun, but we never destroy it. Somewhere, sometime, it will return, purged and fresh, in rain clouds over India or Africa or the rolling hills of Europe. Each day more than 800 million acre-feet of water rains onto the earth. There is, even today, enough to go around. The difficulty is in insuring that water is always where we need it, when we need it, and in a suitable state—for all 7.3 billion of us.

As the late American hydrologist Robert Ambroggi put it almost four decades ago, "The problem facing mankind is not a lack of fresh water, but a lack of efficient regimes for using the water that is available." The fact that we are so bad at managing water at least shows that the potential for doing better is high. The solution in most cases is not more and bigger engineering schemes. It is not south-to-north projects or river-interlinking projects or giant desert canals or megadams. Such supply-side projects are hugely expensive, and often cause as many problems as they solve. They are, I believe, at the heart of our current inefficiency.

To manage the water cycle better, we first have to give up the idea that water has to be extracted from nature and put inside metal or behind concrete before it can be used. We have to treat nature as the ultimate provider of water rather than its wasteful withholder. We must learn to "ride the water cycle" rather than replace it. To treat water as a precious resource rather than something that just falls from the sky.

Technology will play a role. We have to find the "efficient regimes" for using the water cycle that Ambroggi spoke of. That certainly means doing better science and investing in a "blue revolution" to bring the old green-revolution crops in line with hydrological realities. It means "closing the loop" by recycling our sewage, both for its nutrients and its irrigation benefits. It means greater efficiency in our use of water, everywhere from the toilet cistern to the fields. But we must never forget that access to water is a human right. Water cannot always flow toward money, whether up hill or down dale. Efficiency cannot always trump other imperatives, whether on the West Bank or along the Rio Grande.

We need a new ethos for water—an ethos based on managing the water cycle for maximum ecological and social benefit rather than narrow self-interest. That will sometimes mean going back to ancient ways, such as harvesting the rain where it falls. Or adopting the social norms of sharing that often sustained those technologies. There is huge potential for this, and it will often make much more sense than continuing the twentieth-century obsession with large dams and mass transport of water in pipes and canals.

For me, one image of hope is the rainwater harvesters in Gujarat, pouring monsoon water down their wells. Why not develop this concept on a much larger scale? Proposals for the large-scale diversion of the monsoon floodwaters of the Ganges into the aquifer beneath its plain could make sense. The best way might be to pour river water into unlined irrigation canals and let it seep underground, thus turning conventional notions of irrigation efficiency on their head.[7]

We must realize too that water has to be given back to nature. The environmentalists' case for insuring "conservation flows" in rivers and on wetlands is unanswerable. Nature's free services in maintaining fisheries, protecting against floods and drought, cleaning pollution, delivering free irrigation on floodplains, watering valuable tourist sites, and much else are just too valuable to be lost. To insure our water supplies and protect ourselves against damaging floods, especially in an uncertain world of changing climate, we will often have to tear down dams and dikes, recharge the underground reservoirs, and remake the rivers.

The twentieth-century view that the world can feed itself only by artificial irrigation of huge areas of the developing world will also have to go. It is hubris we cannot afford. This kind of irrigation has always been a fantastically expensive and inefficient way of growing crops. That is why bodies like the World Bank have virtually stopped funding most irrigation projects. They should be a last resort. We seem to have forgotten that direct rainfall onto soil is what makes possible well over half the world's crop production. Plant breeders need to redouble their efforts to design crops that can cope with the inevitable variability of that rainfall.

~~~

Perhaps above all we need a new attitude to water and to rivers. Rivers are sacred in most religions. For Buddhists, the gods live at the center of the universe, where the great rivers of their world—the Ganges, the Indus, and the Brahmaputra—rise. Hindus go on pilgrimage to drink the holy waters of the Ganges, and—hundreds of miles downstream, thankfully—they cast the bodies of their dead into the same river. Christianity depicts humanity beginning in the Garden of Eden, where fountains fed rivers that watered the world. In recognition, Christians baptize the faithful in water. Islamic Sharia law was originally a water law. We sing of the river Jordan and the waters of Babylon. The sources of water are everywhere revered. Australian Aborigines hold the waterholes and billabongs of the outback to be sacred places, physical manifestations of the process of creation. The Japanese mark the start of rivers with Shinto temples. Africans pray at springs in the depths of their sacred groves. Europe is peppered with ancient holy wells.

We are inclined to think of rivers as living beings, almost as human. Recently, national governments and courts have begun to reflect this. In 2008, Ecuador gave human-style natural rights to rivers. They had the "right to exist, persist, maintain and regenerate," and three years later environmentalists successfully used the law to defend the river Vilcabamba against harm during road construction.[8]

In early 2017, a court in northern India gave the Ganges and its tributary the Yamuna, which are sacred in the Hindu religion, the status of "living human entities," in an apparent attempt to raise the priority for cleaning them up.[9] The ruling meant that humans could act as legal guardians for rivers and bring court action against polluters. A week before, New Zealand had granted the same legal status to its third-longest river, the Whanganui, following a long-standing demand of a local Maori community, which regarded the river as an ancestor.[10] Again, human guardians will now be able to act for the river in court proceedings. The river, perhaps by way of response, burst its banks a few days later after torrential rains.[11]

Rivers are symbols of nationhood too. Like nations themselves, rivers are always present but always changing. Ol' Man River, he just keeps rolling along. The Yellow River is China's joy and sorrow. Moses parted the Nile to save his people. Russian nationalists in the 1980s rose in anger at the prospect of delivering Siberian water to their fellow socialist republics in Central Asia. The Thames in England and the Murray in Australia, the Loire in France, the Rhine in Germany, and the Mississippi in the United States; trout streams and salmon rivers and even crocodile swamps—all tug at the heartstrings of nations.

This sense of national identity tied up with rivers is part of why we talk these days about going to war over water. Rivers can be sources of reconciliation too, however, and a recognition of codependence. They are too precious, too vital, and too damn slippery to fight over. I was reminded of this at a remarkable interfaith event in 2014 in London, that most multicultural of cities. It began when I met up with a Jewish rabbi at an Indian restaurant in North London. We talked over papadoms and thali before taking a short walk through the streets of Wembley to a local mosque. There we were greeted warmly at the door by our Muslim hosts. In a side room, we joined a clutch of Christians who watched quizzically as the rabbi led Jews in Sabbath prayers amid the accoutrements of a Muslim place of worship.

Then the meeting proper began. People from three faiths and none had gathered to discuss the holiness of water and how to use their shared concern for the elixir of life to rehabilitate a Middle Eastern river reduced to a trickle of sewage effluent. That pathetic, putrid trickle, hidden behind military cordons, is of course the Jordan, a river sacred to half of humanity—Christians, Jews, and Muslims alike—and yet abused by all.

As we met that balmy April evening, the news media was reporting the collapse of the latest peace talks between Israel and Palestine. The assembly was depressed by this but unsurprised. The hundred-plus participants wondered if ecological concerns about the river might help break the political deadlock. Whether a desire to bring the river back to life, by releasing water from upstream dams, could be the trigger for a wider rapprochement. Could hydrology and spirituality succeed where politics had failed?

The 2014 meeting had been instigated by two North London religious leaders. Frank Dabba Smith, a voluble American rabbi who heads the Mosaic Liberal Synagogue, said: "The water professionals can lead the way to peace in the Middle East." Shahab Hussein, an urbane secretary-general of the Wembley mosque, said that Sharia law included the idea that "you can't refuse to give water, especially to a traveler."

To date, the spiritual aspect of water has neither stopped the river Jordan from becoming a hostage to conflict nor prevented its baptismal waters turning to sewage. But the leaders continued to hope. The unlikely pair had first met while giving talks at a nearby London police station. In briefing police about their respective communities, they found themselves in joint dialogue. That had led to efforts to bring together their two traditionally hostile urban communities. And now they had a far wider agenda: peace in the Middle East itself.

They told the meeting participants that the key to peace on the river Jordan—as in North London—lay in interfaith meetings where science, politics, and shared spirituality could come together. It was an experiment. A faith lab, if you will. For many, religion seems to be the cause of conflicts around the world. For Smith and Hussein, it may be the key to defusing conflicts.

The first Wembley interfaith meeting to discuss the river Jordan had taken place two years before, and had been repeated among communities in other places. New York had tried the same thing. The rabbi and the mosque secretary had traveled together to the Middle East, where they linked up with another remarkable body pushing the dream of ecological peace in the Jordan Valley: EcoPeace (formerly Friends of the Earth Middle East), whose Bethlehem boss, water engineer Nader Al-Khateeb, joined them at the Wembley event. Eco-Peace also has offices in Tel Aviv and Amman in Jordan. It was drawing up three national plans—Palestinian, Jordanian, and Israeli—for managing the river. It hoped governments would one day adopt and then merge the plans.

Al-Khateeb told the meeting that an average Palestinian got access to 18 gallons of water a day, while an average Israeli got 80 gallons. The country of Jordan sees virtually none of the river that bears its name. But all must share responsibility, said Munqeth Mehyar, a planning engineer and EcoPeace's Jordanian director. "Yes, Israel took a lot, but Syria dammed the Yarmouk, a tributary, and Jordan has blocked side wadis. We want to get away from the blame game."

The Wembley meeting participants agreed that there were technical reasons why the chances for reviving the forgotten river might be improving. Now that Israel got much of its drinking water from desalination, and recycled much of its sewage effluent, its sense of being under hydrological siege was dissipating. EcoPeace wanted to revive the river Jordan, with half of the new flow coming from Israel and the rest from Syria and Jordan. If all contributed, all could share the benefits of a revived river. But it was more than that, said EcoPeace's founder and director Gidon Bromberg. "If we get it right on water, then we can get it right on refugees and Jerusalem and the settlements and the rest."

It was an inspiring ambition. And one that goes to the heart of a new ethic that recognizes that rivers are more than just sources of water—more than the feedstock for irrigation canals and hydroelectric power stations. It recognizes that rivers provide fish and silt and recharge for underground reserves; that water purges and purifies; that there is virtue in flood pulses and in the mixing of land and water on a river's floodplain. It requires us to find ways of storing water without wrecking the environment, of restoring water to rivers and refilling lakes and wetlands without leaving people thirsty, and of sharing waters rather than fighting over them. It requires us to go with the flow, and to do it before the rivers finally run dry.

ACKNOWLEDGMENTS

THE GESTATION OF THIS BOOK goes back many years, during which time many hundreds of people have helped in its development and in the journalistic research that underpins it. There are too many to name, even if I could remember them all. But I would like to thank those whose assistance brought it to fruition both in its first form, in 2006, and now in this heavily revised and extended second edition more than a decade later. Because I am by training a journalist, I habitually name my sources. So many are quoted or otherwise acknowledged in the text. I hope that will suffice. But some require special mention for guiding me in directions I would never have thought to go, both in my thinking and in my travels.

The first edition of this book might not have happened without the support of Patrick Fuller and the staff of the International Water Management Institute, in Colombo, Sri Lanka. They came up with a travel grant that got me to China and Uzbekistan, supplementing a travel prize for journalism that I won from the CGIAR. In India, I learned a lot from Tushaar Shah and his staff at IWMI, in Anand, Gujarat. IWMI's Iskandar Abdullaev drove me the length of Uzbekistan and introduced me to many stoic people living round the Aral Sea. (My first eye-opening visit to the Aral Sea was thanks to Don Hinrichsen, then at the UN Development Programme.) This second edition has additionally benefited from the IWMI-led Water, Land, and Ecosystems project, which commissioned blogs and provided insights and occasional travel funding.

In this second edition, Jane Madgwick at Wetlands International helped immensely, in particular getting me to Mali and Senegal, where her lieutenants Bakary Kone and Papa Mawade Wade were gurus of their own. Oxfam's Geoff Graves took me to water-starved villages in Palestine, while Arie Issar at the Ben-Gurion University of the Negev, Gidon Bromberg at EcoPeace, and London rabbi Frank Dabba Smith all underlined for me how there is decency to be found on both sides of the water divide there. In Bangladesh and West Bengal, Dipankar Chakraborti was invaluable in first unravelling the mysteries of the arsenic disaster. Eric Baran of WorldFish took me up the Tonle Sap in

Cambodia. Mary Kelly of Environmental Defense guided me well on the Rio Grande. And in China, Xue Yungpeng and many others at the Yellow River Conservancy Commission proved candid and optimistic on a long and fruitful journey down their river.

My travels in Oman were helped by Abdullah Al-Ghafri and Slim Zekri. Before the Syrian civil war, ICARDA in Aleppo gave me the chance to visit *qanats* there. David Riebold invited me to his home in Lanzarote. Bianca Shead, then of SAB Miller, organized my visit to the páramos in Colombia. In Poland, WWF's Przemyslaw Nawrocki showed me the shocking state of its rivers. I visited Honduras twice, once to survey the damage of Hurricane Mitch for the Red Cross's *World Disasters Report* and again, with the assistance of the Pulitzer Center on Crisis Reporting, to investigate the assassination of anti-dam campaigner Berta Cáceres. On the second occasion, huge thanks to Raul Valdivia.

Many other NGOs have helped with their expertise. In California, I have been assisted over the years by International Rivers, especially Peter Bosshard and Lori Pottinger, who is now still on the case at the Public Policy Institute of California, and by Peter Gleick at the Pacific Institute. Likewise, the Stockholm International Water Institute's World Water Week in Sweden; the International Institute for Applied Systems Analysis in Austria; the Oxford Water Network; the Fletcher School of Law and Diplomacy at Tufts University near Boston, which invited me to a workshop on water security in 2015; and the European Centre for River Restoration, whose valuable 2014 event in Vienna I attended. I explored water issues in Australia after speaking at the Brisbane International Riversymposium in 2006. Even further back, Janos Vargha, Pat Adams, Dai Qing, and Phil Williams alerted me to the perils of dams.

Beacon Press, my publishers in the US, have been magnificent throughout and were keen to initiate this second edition. Thanks especially to Amy Caldwell and, for this edition, to Will Myers. In the UK, Susanna Wadeson at Transworld was the first to commission the book. Thanks to Laura Barber and colleagues at Portobello Books for picking up the baton for the second edition. Special thanks also to Jessica Woollard, my agent, who first approached me to write a book about water more than fifteen years ago.

Unknowingly, editors for numerous magazines, newspapers, and websites have helped my travels by paying fares and commissioning articles. Notable among them are Bill O'Neill during his time at the *Guardian*, Roger Cohn at *Yale e360*, and the many editors I have worked with at *New Scientist* over more than thirty-five years. To them all, thanks again.

NOTES

1
The Human Sponge

1. Water Footprint Network, http://waterfootprint.org/en/resources/water-footprint
-statistics.
2. Fernando Jaramillo and Georgia Destouni, "Local Flow Regulation and Irrigation Raise Global Human Water Consumption and Footprint," *Science* 350 (2015):
1248–51.
3. Ertug Ercin, Daniel Chico Zamanillo, and Ashok Chapagain, *Dependencies
of Europe's Economy on Other Parts of the World in Terms of Water Resources*,
Horizon2020-Imprex Project, Technical Report D12.1 (2016), Water Footprint
Network, http://waterfootprint.org/media/downloads/Imprex-D12-1_final.pdf.
4. John A. Allan, *The Middle East Water Question: Hydropolitics and the Global
Economy* (New York: Tauris, 2001); Tony Allan, "'Virtual Water': A Long-Term
Solution for Water-Short Middle Eastern Economies?" working paper, SOAS,
University of London, 1997, https://www.soas.ac.uk/water/publications/papers
/file38347.pdf.
5. Charlotte de Fraiture et al., *Does International Cereal Trade Save Water?: The Im-
pact of Virtual Water Trade on Global Water Use*, Research Report 4 (Colombo,
Sri Lanka: IWMI, 2004), http://ageconsearch.umn.edu/bitstream/92832/2
/CARR4.pdf.

2
Lake Chad

1. George E. Hollis et al., *The Hadejia-Nguru Wetlands: Environment, Economy and
Sustainable Development of a Sahelian Floodplain Wetland* (Gland, Switzerland:
IUCN, 1993).
2. Lekan Oyebande, "Streamflow Regime Change and Ecological Response in the
Lake Chad Basin in Nigeria," in *Hydro-ecology: Linking Hydrology and Aquatic
Ecology*, workshop proceedings, publication no. 266 (London: IAHS Press, 2001).
3. Ahmad Salkida, "Africa's Vanishing Lake Chad," *Africa Renewal*, April 2012, http://
www.un.org/africarenewal/magazine/april-2012/africa%E2%80%99s-vanishing
-lake-chad.
4. *Joint Environmental Audit on the Drying Up of Lake Chad* (Pretoria: GIZ, 2015),
https://www.giz.de/de/downloads/giz2015-en-joint-environmental-audit-report
-lake-chad.pdf.

5. Edward Barbier, "Wetlands as Natural Assets," *Hydrological Sciences Journal* 56 (2011): 1360–73.
6. Caterina Batello, Marzio Marzot, and Adamou Harouna Touré, *The Future Is an Ancient Lake: Traditional Knowledge, Biodiversity and Genetic Resources for Food and Agriculture in Lake Chad Basin Ecosystems* (Rome: FAO, 2004), http://www .fao.org/docrep/010/y5118e/y5118e00.htm.
7. *Joint Environmental Audit on the Drying Up of Lake Chad.*
8. "In the Lake Chad Basin, Populations Are Trapped Between Climate Change and Insecurity," ACTED, December 17, 2015, http://www.acted.org/fr/node/12088.
9. Churchill Okonkwo, "Characteristics of Drought Indices and Rainfall in Lake Chad Basin," *International Journal of Remote Sensing* 34 (2013): 7945–61.
10. *Joint Environmental Audit on the Drying Up of Lake Chad.*
11. Ibid.
12. Lake Chad Basin Commission, *The Lake Chad Development and Climate Resilience Action Plan* (Washington, DC: World Bank, 2016), https://openknowledge .worldbank.org/handle/10986/23793.
13. *Environmental Audit on the Drying Up of the Lake Chad: A Focus on Water Resources Quantity Management by the Nigerian Government: 2008–2013* (Pretoria: GIZ, 2015), http://afrosai-e.intohost.co.za/uploads/afrosai_intohost_co_za/cms /files/environmental_audit_on_the_drying_up_of_lake_chad_nigeria.pdf.
14. "Beyond Chibok," UNICEF, 2016, https://www.unicef.org/infobycountry/files /Beyond_Chibok.pdf.
15. Moki Kindzeka, "Lake Chad Recedes to Catastrophic Levels," Deutsche Welle, November 27, 2015, http://www.dw.com/en/lake-chad-recedes-to-catastrophic -levels/a-18879406.
16. Lachlan Carmichael, "Africa's Lake Chad Could Fuel New Migrant Crisis: UN," AFP, November 8, 2015, https://www.yahoo.com/news/africas-lake-chad-could -fuel-migrant-crisis-un-044107554.html.
17. Thomas Fessy, "Boko Haram Attack: What Happened in Baga?" *BBC News*, February 2, 2015, http://www.bbc.co.uk/news/world-africa-30987043.
18. Mark Piggott, "Bodies of 42 Fishermen Butchered by Boko Haram Pulled from Lake Chad," *International Business Times*, June 14, 2016, http://www.ibtimes.co.uk /bodies-42-fishermen-butchered-by-boko-haram-pulled-lake-chad-1565498.

3
Sahel

1. Aziz Salmone, "Conflict in the Senegal River Valley," *Cultural Survival Quarterly* (December 1998), https://www.culturalsurvival.org/publications/cultural-survival -quarterly/conflict-senegal-river-valley.
2. Senegal River Development Organization, *Transboundary Diagnostic Environmental Analysis of the Senegal River Basin* (Washington, DC: GEF, 2014), available from International Waters Learning Exchanges & Research Network, http://iwlearn.net /resources/documents/2705; Fred Pearce, "The Myth of the 'Win-Win,'" *Thrive* (blog), CGIAR, July 14, 2016, https://wle.cgiar.org/thrive/2016/07/14/myth-win-win.
3. Michael C. Acreman and G. E. Hollis, *Water Management and Wetlands in Sub-Saharan Africa* (Gland, Switzerland: IUCN, 1996).

4. M. Tayaa et al., *Canary Current: GIWA Regional Assessment 41* (Kalmar, Sweden: University of Kalmar, on behalf of UNEP, 2005), http://citeseerx.ist.psu.edu /viewdoc/download?doi=10.1.1.398.5337&rep=rep1&type=pdf.

5. Kindzeka, "Lake Chad Recedes to Catastrophic Levels."

6. Susanna Davies, *Adaptable Livelihoods: Coping with Food Insecurity in the Malian Sahel* (New York: St. Martin's, 1996).

7. Leo Zwarts et al., *The Niger, a Lifeline* (Wageningen, Netherlands: Wetlands International, 2005), https://www.wetlands.org/publications/the-niger-a-lifeline.

8. *The Mali Migration Crisis at a Glance* (Geneva: International Organization for Migration, 2013), https://www.iom.int/files/live/sites/iom/files/Country/docs/Mali _Migration_Crisis_2013.pdf.

9. "Mali: Lawlessness, Abuses Imperil Population," Human Rights Watch, press release, April 14, 2015, https://www.hrw.org/news/2015/04/14/mali-lawlessness -abuses-imperil-population.

10. Drew Hinshaw and Mackenzie Knowles-Coursin, "'African Dream' of Europe Turns to Nightmare," *Wall Street Journal*, December 29, 2015, https://www.wsj .com/articles/african-dream-of-europe-turns-into-a-nightmare-1451405269; Alex Duval Smith, "From Mali to Sicily," *BBC News*, April 21, 2015, http://www.bbc.co .uk/news/world-africa-32392788.

11. L. Zwarts and J.-L. Frerotte, *Water Crisis in the Inner Niger Delta (Mali): Causes, Consequences, Solutions*, report no. 1832 (Feanwâlden, Netherlands: Altenburg & Wymenga, 2012), http://www.altwym.nl/uploads/file/490_1369389376.pdf.

4
Riding the Water Cycle

1. Peter Gleick, ed., *Water in Crisis: A Guide to the World's Fresh Water Resources* (New York: Oxford University Press, 1993).

2. Scott Jasechko et al., "Global Aquifers Dominated by Fossil Groundwaters but Wells Vulnerable to Modern Contamination," *Nature Geoscience* 10 (2017): 425–29.

3. Paul Ehrlich, *The Population Bomb* (London: Pan/Ballantine, 1971).

4. Aidan Gulliver et al., *Farming Systems and Poverty: Improving Farmers' Livelihoods in a Changing World* (Rome: FAO, 2001).

5. Meredith Giordano et al., eds., *Water for Wealth and Food Security: Supporting Farmer-Driven Investments in Agricultural Water Management* (Colombo, Sri Lanka: IWMI, 2012), http://www.iwmi.cgiar.org/Publications/Other/Reports/PDF /Water_for_wealth_and_food_security.pdf.

6. Charlotte de Fraiture et al., "Small Pumps and Poor Farmers in Sub-Saharan Africa," *Water International* 38 (2013): 827–39.

7. Arjen Hoekstra et al., "Global Monthly Water Scarcity: Blue Water Footprints Versus Blue Water Availability," *PLOS One*, February 29, 2012, http://journals.plos .org/plosone/article?id=10.1371/journal.pone.0032688.

8. Mesfin Mekonnen and Arjen Hoekstra, "Four Billion People Facing Severe Water Scarcity," *Science Advances*, February 12, 2016, http://advances.sciencemag.org /content/2/2/e1500323.

9. Manzoor Qadir et al., "Economics of Salt-Induced Land Degradation and Restoration," *Natural Resources Forum* 38 (2014): 282–95.

5
Rio Grande

1. Lorne Matalon, "Border Wetland Uses Treated Wastewater as Congress Considers Wetland Funding," Fronteras Desk, May 26, 2015, http://www.fronterasdesk.org /content/10032/border-wetland-uses-treated-wastewater-congress-considers -wetland-funding.
2. William Eichinger et al., *Lake Evaporation Estimation in Arid Environments*, IIHR report no. 430 (Iowa City: University of Iowa, 2003), http://www.iihr.uiowa.edu /wp-content/uploads/2013/06/IIHR430.pdf.
3. "Past and Present Water Supplies," El Paso Water, http://www.epwu.org/water /water_resources.html.
4. Ashley Madonna and Abbie Maynard, "Running Water and Progress Come Slowly to Texas Colonias," *Austin American-Statesman*, May 21, 2016, http://www .mystatesman.com/news/running-water-and-progress-come-slowly-texas-colonias /adyLz4LJoPQ5x0Qk1gqv4N.
5. "Mexico Pays Rio Grande Water Debt in Full," US Embassy Mexico, press release, February 26, 2016, https://mx.usembassy.gov/mexico-pays-rio-grande-water-debt-in-full.
6. Mary Kelly and Gerardo Jiménez González, *The Ojinaga Valley: At the Confluence of the Lower Río Conchos and the Río Bravo* (New York: Environmental Defense Fund, 2004), http://www.edf.org/sites/default/files/2_OjinagaValley.pdf.
7. Concepcion Luján-Alvarez and Mary Kelly, *Agricultural Irrigation Conservation Projects in the Delicias, Chihuahua Irrigation District: A Report on Public Participation, Certification and Early Implementation* (New York: Environmental Defense Fund, 2012), http://bva.colech.edu.mx/xmlui/handle/123456789/HASH8258of4475 c8e821ed1f4d; "Agricultural Irrigation Conservation Projects in the Delicias, Chihuahua Irrigation District," NADB, http://www.nadbank.org/pdfs/state_projects /FS%20Delicias%20Irrigation%2010-02%20_Eng_.pdf.

6
Colorado

1. Brian Clark Howard, "Historic 'Pulse Flow' Brings Water to Parched Colorado River Delta," *National Geographic*, March 24, 2014, http://news.nationalgeographic.com /news/2014/03/140322-colorado-river-delta-pulse-flow-morelos-dam-minute-319-water.
2. Eloise Kendy, "Colorado River: Six Months After the Pulse Flow," Nature Conservancy, 2014, https://www.nature.org/ourinitiatives/regions/northamerica/areas /coloradoriver/colorado-river-six-months-after-the-pulse-flow.xml.
3. Marc Reisner, *Cadillac Desert: The American West and Its Disappearing Water* (New York: Viking, 1986).
4. Bradley Udall and Jonathan Overpeck, "The Twenty-First-Century Colorado River Hot Drought and Implications for the Future," *Water Resources Research* 53, no. 3 (2017), http://onlinelibrary.wiley.com/doi/10.1002/2016WR019638/pdf.
5. "Lake Mead Water Level," Lakes Online, http://mead.uslakes.info/level.asp.
6. Julie Vano et al., "Understanding Uncertainties in Future Colorado River Streamflow," *BAMS*, January 2014, http://journals.ametsoc.org/doi/pdf/10.1175/BAMS-D -12-00228.1.

7. Brian Richter et al., "Tapped Out: How Can Cities Secure Their Water Future?" *Water Policy* 15 (2013): 335–63.

8. Western Resource Advocates, *Arizona's Water Future: Colorado River Shortage, Innovative Solutions, and Living Well with Less*, 2017, http://westernresource advocates.org/wp-content/uploads/2017/01/WRA-Arizonas-Water-Future-Exec-Summary-Report-vfinal.pdf.

9. "Colorado River Water Supply Report System Contents," November 28, 2017, Central Arizona Project, https://www.cap-az.com/documents/planning/CRreport SupplyReport.pdf.

10. US Bureau of Reclamation, Lower Colorado Region, Yuma Area Office, https://www.usbr.gov/lc/yuma/facilities/ydp/yao_ydp.html, accessed 2017.

7
California

1. "California Must Ration Water to Avoid Drought Disaster," *New Scientist*, March 18, 2015, https://www.newscientist.com/article/mg22530133-900.

2. Adam Nagourney, "California Imposes First Mandatory Water Restrictions to Deal with Drought," *New York Times*, April 1, 2015, https://www.nytimes.com /2015/04/02/us/california-imposes-first-ever-water-restrictions-to-deal-with -drought.html.

3. Paul Rogers, "How the Drought Changed California Forever," *Mercury News* (San Jose, CA), April 17, 2017, http://www.mercurynews.com/2017/04/15/how-the -drought-changed-california-forever.

4. Deepti Singh et al., *An Increase in Extreme-Weather Winters for the United States*, Stanford Woods Institute for the Environment, research brief, Fall 2016, https:// woods.stanford.edu/sites/default/files/files/Extreme-Weather-Winters-Research -Brief.pdf.

5. "The Great Flood," Greetings from the Salton Sea, http://www.greetingsfrom saltonsea.com/flood.html, accessed 2017.

6. Eric Niiler, "Salton Sea Resort Area Hoping for Return of Glory Days," *San Diego Union-Tribune*, June 30, 1998, http://www.sci.sdsu.edu/salton/NiilerSaltonSea Resort.html.

7. "Salton Sea," Pacific Institute, http://pacinst.org/issues/sustainable-water -management-local-to-global/salton-sea, accessed 2017.

8. Michael J. Cohen, *Hazard's Toll: The Costs of Inaction at the Salton Sea* (Oakland, CA: Pacific Institute, 2014), http://pacinst.org/app/uploads/2014/09/PacInst _HazardsToll.pdf.

9. Edwin Delgado, "Salton Sea Plan Features Goal of Conservation," *Imperial Valley Press* (El Centro, CA), May 26, 2017, https://www.pressreader.com/usa/imperial -valley-press/20170326/281479276245368.

8
Mekong

1. Larry Lohmann, *Mekong Dams in the Drama of Development* (Dorset, UK: Corner House, 1998), http://www.thecornerhouse.org.uk/resource/mekong-dams-drama -development.

2. *Independent Expert Review of the Pak Beng Dam Environmental Impact Assessment and Supporting Project Documents* (Berkeley, CA: International Rivers, May 2017), https://www.internationalrivers.org/sites/default/files/attached-files/independent expertreview_pakbengdameia_may2017.pdf.

3. "Mekong Dams Concern UN Researchers," Chinadialogue, 2009, https://www.chinadialogue.net/blog/3030-Mekong-dams-concern-UN-researchers-/en.

4. Fred Pearce, "The Damming of the Mekong: Major Blow to an Epic River," *Yale e360*, June 16, 2009, http://e360.yale.edu/features/the_damming_of_the_mekong_major_blow_to_an_epic_river.

5. Francois Molle et al., *Contested Waterscapes in the Mekong Region: Hydropower, Livelihoods and Governance* (London: Earthscan, 2009).

6. Colin Poole, *Tonle Sap: The Heart of Cambodia's Natural Heritage* (Bangkok: River Books, 2005).

7. D. Nettleton and Eric Baran; "Fishery Stakeholder Groups and Livelihood Variation Around the Tonle Sap Great Lake Cambodia," WorldFish, 2003 (mimeograph); Duc San Tana, *The Inland and Marine Fisheries Trade of Cambodia* (Phnom Penh, Cambodia: Oxfam America, 2003).

8. Kathy Marks, "Low Plains Drifters," *Independent*, June 19, 2004, http://www.independent.co.uk/travel/asia/low-plains-drifters-732887.html.

9. Yong Zhong et al., "Rivers and Reciprocity: Perceptions and Policy on International Watercourses," *Water Policy* (2016), http://wp.iwaponline.com/content/early/2016/02/29/wp.2016.229.

10. Juha Sarkkula et al., "Ecosystem Processes of the Tonle Sap Lake," working paper, 2003, http://citeseerx.ist.psu.edu/viewdoc/summary?doi=10.1.1.472.3903.

11. Eric Baran et al., "Floods, Floodplains and Fish Production in the Mekong Basin: Present and Past Trends," in *Proceedings of the Second Asian Wetlands Symposium*, Penang, Malaysia, August 27–30, 2001, http://pubs.iclarm.net/resource_centre/WF_1003.pdf.

12. Yong Zhong et al., "Rivers and Reciprocity: Perceptions and Policy on International Watercourses," *Water Policy* (2016), doi: 10.2166/wp.2016.229.

13. Xuezhong Yu, "Transboundary Hydropower Operation Calls for Long-Term Mechanism in the Lancang-Mekong River Basin," *Mekong Blog*, CGIAR, April 19, 2016, https://wle-mekong.cgiar.org/transboundary-hydropower-operation-calls-for-long-term-mechanism.

9

Seas of Death

1. Sean Avery, *What Future for Lake Turkana?*, African Studies Centre, University of Oxford, 2013, http://www.africanstudies.ox.ac.uk/sites/sias/files/documents/WhatFutureLakeTurkana-%20update.pdf.

2. "Ethiopia: Dams, Plantations a Threat to Kenyans," Human Rights Watch, 2017, https://www.hrw.org/news/2017/02/14/ethiopia-dams-plantations-threat-kenyans.

3. *"What Will Happen If Hunger Comes?": Abuses Against the Indigenous Peoples of Ethiopia's Lower Omo Valley* (New York: Human Rights Watch, 2012), https://www.hrw.org/sites/default/files/reports/ethiopia0612webwcover_0.pdf.

4. Kate Ravilious, "Many of the World's Lakes Are Vanishing," *New Scientist*, March 4, 2016, https://www.newscientist.com/article/2079562.

5. Somayeh Shadkam et al., "Preserving the World's Second Largest Hypersaline Lake," *Science of the Total Environment* 559 (2016): 317–25; Ali Mirchi et al., "Lake Urmia: How Iran's Most Famous Lake Is Disappearing," *Guardian*, January 23, 2015, https://www.theguardian.com/world/iran-blog/2015/jan/23/iran-lake-urmia -drying-up-new-research-scientists-urge-action.

6. Edward Maltby, *Waterlogged Wealth* (London: Earthscan, 1986); Jeremy Purseglove, *Taming the Flood* (New York: Oxford University Press, 1988).

7. Max Finlayson, *State of Global Wetlands and Implications for the Sustainable Development Goals* (New South Wales, Australia: Institute for Land, Water, & Society, 2015), http://riversymposium.com/wp-content/uploads/2015/10/Dr-Max-Finlayson.pdf.

8. Nina Strochlic, "Massacre Survivors Cling to Life in Giant Swamp," *National Geographic*, June 14, 2017, http://news.nationalgeographic.com/2017/02/sudd-south-sudan.

9. Andrew Balmford et al., "Economic Reasons for Conserving Wild Nature," *Science* 297 (2002): 950–53, http://ucsb.piscoweb.org/~teck/EEMB168/Balmford_etal _2002.pdf.

10. "Iran-Afghanistan Talks Focused on Helmand Water Right," *Mehr News*, October 26, 2015, http://en.mehrnews.com/news/111377/Iran-Afghanistan-talks-focused-on -Helmand-water-right.

11. Najmeh Bozorgmehr, "Ecological Disaster Looms in Iran's Dying Wetlands," *Financial Times*, February 13, 2015, https://www.ft.com/content/8c555072-ac6e-11e4 -9d32-00144feab7de.

12. Azizullah Hamdard, "Kajaki Dam Expansion Contract Signed with Turkish Firm," *Pajhwok News*, October 1, 2016, http://www.pajhwok.com/en/2016/10/01/kajaki -dam-expansion-contract-signed-turkish-firm; Sune Engel Rasmussen, "Dam Project Promises Water—as Well as Conflict," *Guardian*, March 22, 2017, https:// www.theguardian.com/global-development/2017/mar/22/afghanistan-dam -project-iran-nimruz-helmand-river.

10
England

1. "Chalk Streams," Wildlife Trusts, http://www.wildlifetrusts.org/wildlife/habitats /chalk-streams, accessed May 2017.

2. *The State of England's Chalk Rivers* (Rotherham, UK: Environment Agency, 2004), http://adlib.everysite.co.uk/resources/000/057/268/Summary_chalk_rivers.pdf.

3. "Calls for Better Protection for the Itchen," Wildlife Trusts, January 17, 2014, http://www.hiwwt.org.uk/news/2014/02/27/calls-better-protection-itchen-0.

4. "Over Half of Chalk Streams and a Quarter of Rivers in England Currently at Risk Due to Poor Water Management and Usage," WWF, June 25, 2017, https://www .wwf.org.uk/updates/over-half-chalk-streams-and-quarter-rivers-england -currently-risk-due-poor-water-management.

11
India

1. Aditi Deb Roy and Tushaar Shah, "The Socio-Ecology of Groundwater in India," IWMI-Tata Water Policy Program, *Water Policy Briefing* 4 (2002), http://www .iwmi.cgiar.org/Publications/Water_Policy_Briefs/PDF/wpb04.pdf.

2. "Changing Rainfall Patterns Linked to Water Security in India," IIASA, January 9, 2017, http://www.iiasa.ac.at/web/home/about/news/170109-groundwater.html.

3. Jenny Grönwall, "Power to Segregate: Improving Electricity Access and Reducing Demand in Rural India," working paper, SIWI, 2014, http://www.siwi.org /publications/power-to-segregate-improving-electricity-access-and-reducing -resource-demand-in-rural-india.

4. Aditi Mukherji and Tushaar Shah, "Groundwater Governance in South Asia: Governing a Colossal Anarchy," IWMI-TATA, Water Policy Research Highlight 13, http://www.iwmi.cgiar.org/iwmi-tata_html/PM2003/PDF/13_Highlight.pdf.

5. Tushaar Shah et al., "Sustaining Asia's Groundwater Boom," Natural Resources Forum (2003), doi: 10.1111/1477-8947.00048.

6. Giordano et al., *Water for Wealth and Food Security*; Christopher Scott and Majed Akhter, review of Tushaar Shah's *Taming the Anarchy: Groundwater Governance in South Asia* (2009), *Human Ecology* (2010), http://udallcenter.arizona.edu/wrpg /pubs2010/Scott%20&%20Akhter%202010%20Shah%20GW%20anarchy%20 review%20HE.pdf.

7. Victor Mallet, "India: Water Wars," *Financial Times*, April 13, 2016, https://www .ft.com/content/96687242-009b-11e6-ac98-3c15a1aa2e62.

12

Overpumping the World

1. John McNeill, *Something New Under the Sun: An Environmental History of the Twentieth-Century World* (London: Allen Lane, 2000).

2. Sandra Postel and Chris Martenson, "Repairing the Water Cycle," interview, *Resilience*, February 11, 2016, http://www.resilience.org/stories/2016-02-11/sandra-postel -repairing-the-water-cycle.

3. Carole Dalin et al., "Groundwater Depletion Embedded in International Food Trade," *Nature* 543 (2017): 700–704.

4. "Groundwater Governance in the Arab World," IWMI, http://gw-mena.iwmi.org, accessed 2017.

5. Karen Villholth and Yvan Altchenko, "Is Groundwater the Key to Increasing Food Security in Sub-Saharan Africa?" CGIAR, *Thrive* (blog), April 23, 2016, https:// wle.cgiar.org/thrive/2016/04/23/groundwater-key-increasing-food-security-sub -saharan-africa.

6. Michael Marshall, "Vast Supplies of Groundwater Found Under Kenya," *New Scientist*, September 10, 2013, https://www.newscientist.com/article/dn24177.

7. *The Great Man-Made River Project* (Tripoli: Socialist People's Libyan Arab Jamahiriya, 1989).

8. Raymond Bonner, "Libya's Vast Desert Pipeline Could Be Conduit for Troops," *New York Times*, December 2, 1997, http://www.nytimes.com/1997/12/02/world /libya-s-vast-desert-pipeline-could-be-conduit-for-troops.html.

9. Nafeez Ahmed, "War Crime: NATO Deliberately Destroyed Libya's Water Infrastructure," *Ecologist*, May 14, 2015, http://www.theecologist.org/News/news_analysis /2869234/war_crime_nato_deliberately_destroyed_libyas_water_infrastructure.html.

10. Kieran Cooke, "Trouble Ahead for Gaddafi's Great Man-Made River," *Middle East Eye*, December 28, 2016, http://www.middleeasteye.net/columns/trouble-great -man-made-river-1331047422.

11. Elie Elhadj, "Camels Don't Fly; Deserts Don't Bloom," Occasional Paper no. 48, Water Issues Study Group, SOAS, 2004, https://www.soas.ac.uk/water/publications/papers/file38391.pdf.

12. Fred Pearce, *The Landgrabbers: The New Fight over Who Owns the Earth* (Boston: Beacon Press, 2012).

13. Adnan Ghosheh, "Water Situation Alarming in Gaza," World Bank, November 22, 2016, http://www.worldbank.org/en/news/feature/2016/11/22/water-situation-alarming-in-gaza.

14. Jad Isaac, *A Sober Approach to the Water Crisis in the Middle East* (Jerusalem: Applied Research Institute, 1995), https://arij.org/files/admin/1995_A_sober_approach_to_the_water_crisis_in_the_middle_east.pdf.

15. Peter Beaumont, "'The Worst It's Been': Children Continue to Swim as Raw Sewage Floods Gaza Beach," *Guardian*, July 31, 2017, https://www.theguardian.com/cities/2017/jul/31/children-swim-sewage-floods-gaza-beach.

16. Amira Hass, "Gaza Water Crisis Has Caused Irreversible Damage, World Bank Warns," *Haaretz*, December 18, 2016, http://www.haaretz.com/middle-east-news/palestinians/.premium-1.759600.

13
Bangladesh

1. Sarah Flanagan et al., "Arsenic in Tube Well Water in Bangladesh, *WHO Bulletin* (September 2012), http://www.who.int/bulletin/volumes/90/11/11-101253/en.

2. Human Rights Watch, *Nepotism and Neglect: The Failing Response to Arsenic in the Drinking Water of Bangladesh's Rural Poor* (New York: Human Rights Watch, 2016), https://www.hrw.org/report/2016/04/06/nepotism-and-neglect/failing-response-arsenic-drinking-water-bangladeshs-rural.

3. "Bangladesh: 20 Million Drink Arsenic-Laced Water," Human Rights Watch, press release, April 6, 2016, https://www.hrw.org/news/2016/04/06/bangladesh-20-million-drink-arsenic-laced-water.

4. Fred Pearce, "Arsenic in the Water," *Guardian*, February 19 and 26, 1998.

5. Barun Kumar Thakur, "Groundwater Arsenic Contamination in Bihar: Causes, Issues and Challenges," *Manthan*, July 2015, https://www.researchgate.net/publication/281865761_Groundwater_Arsenic_Contamination_in_Bihar_Causes_Issues_and_Challenges.

6. "Groundwater Arsenic Spreading Cancer in Several Bihar Districts," *Times of India*, March 13, 2016, http://timesofindia.indiatimes.com/life-style/health-fitness/health-news/Ground-water-arsenic-spreading-cancer-in-several-Bihar-districts/articleshow/51381338.cms; Medhavi Arora, "Arsenic-Polluted Water Linked to Cancer in India," CNN.com, May 1, 2017, http://edition.cnn.com/2017/04/28/health/arsenic-water-pollution-cancer-india.

7. Manish Kumar et al., "Arsenic Enrichment in Groundwater in the Middle Gangetic Plain of Ghazipur District in Uttar Pradesh, India," *Journal of Geochemical Exploration* 105 (2010): 83–94; Neha Lalchandanil, "Silent Killer Arsenic Poisons Lives in Ballia," *Times of India*, October 31, 2011, http://timesofindia.indiatimes.com/india/Silent-killer-arsenic-poisons-lives-in-Ballia/articleshow/10549386.cms.

8. Arunav Sinhai, "30 UP Districts in Grip of Arsenic Poisoning," *Times of India*, July 14, 2014, http://timesofindia.indiatimes.com/city/lucknow/30-UP-districts-in-grip-of-arsenic-poisoning/articleshow/38339052.cms.

9. Stacy Morford, "Urban Pumping Raises Arsenic Risk in Southeast Asia," Phys.org, August 22, 2016, https://phys.org/news/2016-08-urban-arsenic-southeast-asia.html.

10. Mason Stahl et al., "River Bank Geomorphology Controls Groundwater Arsenic Concentrations in Aquifers Adjacent to the Red River, Hanoi Vietnam," *Water Resources Research* 52 (2016): 6321–34.

11. Luis Rodriguez-Lado et al., "Groundwater Arsenic Contamination Throughout China," *Science* 341 (2013): 866–68.

14
Wonders of the World

1. Bill Gates, "Have You Hugged a Concrete Pillar Today?" *Gatesnotes*, https://www.gatesnotes.com/Books/Making-the-Modern-World.

2. Quoted in Donald Worster, *Under Western Skies: Nature and History in the American West* (New York: Oxford University Press, 1992), 70.

3. Marshall Goldman, *USSR in Crisis: The Failure of an Economic System* (New York: W. W. Norton, 1983).

4. *The New Great Walls: A Guide to China's Overseas Dam Industry*, 2nd ed. (Berkeley, CA: International Rivers, 2012), https://www.internationalrivers.org/sites/default/files/attached-files/intlrivers_newgreatwalls_2012_2.pdf.

5. Daniel Beard, "Creating a Vision of Rivers for the 21st Century," remarks presented at International Dam Summit, Nagaragawa, Japan, September 14, 1996, http://www.bosnia.ba/neretva/slike/gotova%20je%20era%20brana%20-%20eng.txt.

6. David Grey and Claudia Sadoff, "Water for Growth and Development," in *Thematic Documents of the 4th World Water Forum* (Mexico City: Comision Nacional del Agua, 2006), http://siteresources.worldbank.org/INTWRD/Resources/FINAL_0601_SUBMITTED_Water_for_Growth_and_Development.pdf.

7. Peter Bosshard, *Infrastructure for Whom?: A Critique of the Infrastructure Strategies of the Group of 20 and the World Bank* (Berkeley, CA: International Rivers, 2012), https://www.internationalrivers.org/sites/default/files/attached-files/infrastructure_for_whom_report.pdf.

8. Gary Owen, "Lesotho: Famine Fears in African Country Twinned with Wales," *BBC News,* May 12, 2016, http://www.bbc.co.uk/news/uk-wales-36270169.

9. Solomon Kibret et al., "Malaria and Large Dams in Sub-Saharan Africa," *Malaria Journal* 15 (2016): 448.

15
The New Dam Era

1. Jamie Skinner and Lawrence Haas, "Renewed Hydropower Investment Needs Social and Environmental Safeguards," briefing, IIED, 2014, http://pubs.iied.org/17206IIED.

2. Christiane Zarfl et al., "A Global Boom in Hydropower Dam Construction," *Aquatic Sciences* 77 (2015): 161–70.

3. Jonathan Watts, "Belo Monte, Brazil: The Tribes Living in the Shadow of a Megadam," *Guardian*, December 16, 2014, https://www.theguardian.com/environment/2014/dec/16/belo-monte-brazil-tribes-living-in-shadow-megadam.

4. Grace Mang, "China's Global Dam Builders: Talking the Talk, Walking the Walk?," International Rivers, June 23, 2015, https://www.internationalrivers.org /node/9063.

5. "Thousands Flooded by Sudan Dam Closure—Villagers," Reuters, October 2, 2008, https://www.business-humanrights.org/en/thousands-flooded-by-sudan -dam-closure-villagers; Magaji Isa, "3050MW Mambilla Hydro Power Project Set to Start," *Daily Trust*, September 6, 2016, https://www.dailytrust.com.ng/news /general/3050mw-mambilla-hydro-power-project-set-to-start/161544.html; Jona- than Watts, "Argentina Leader Leaves Controversial Legacy with Patagonia Dams Project," *Guardian*, December 1, 2015, https://www.theguardian.com/world/2015 /dec/01/argentina-president-cristina-fernandez-de-kirchner-patagonia-hydroelectric -dam-project.

6. Fred Pearce, "Zambezi: The Mekong's Diplomatic Debacle on Repeat," *Thrive* (blog), CGIAR, September 9, 2013, https://wle.cgiar.org/thrive/2013/09/12/zambezi -mekongs-diplomatic-debacle-repeat.

7. Ibid.

8. Gregory Poindexter, "AfDB Named Lead Coordinator for 2,400-MW Batoka Gorge Hydropower Project in Africa," *Hydroworld*, April 18, 2017, http://www .hydroworld.com/articles/2017/04/afdb-named-lead-coordinator-for-2-400-mw -batoka-gorge-hydropower-project-in-africa.html.

9. David Rogers, "Chinese Bidder Praised for Speed in DR Congo's Massive Dam Con- test," *Global Construction Review* (May 9, 2016), http://www.globalconstruction review.com/news/chinese-bidder-praised-speed-d7r-co7ngos-mas7sive/.

10. *IDA Support to Transformational Projects with Regional Impact* (Washington, DC: IDA, 2013), http://documents.worldbank.org/curated/en/710721468158072319/pdf /759270BR0IDA0S00Disclosed0308020130.pdf.

11. "Congo's Energy Divide," International Rivers, June 9, 2013, https://www .internationalrivers.org/resources/congo's-energy-divide-factsheet-3413.

12. Atif Ansar et al, "Should We Build More Large Dams?," *Energy Policy* 69 (2014): 43–56.

13. Ted Veldkamp, "Water Scarcity Hotspots Travel Downstream Due to Human Interventions in the 20th and 21st Century," *Nature Communications* (2017), doi: 10.1038/ncomms15697.

14. Peter Greste, "The Dam That Divides Ethiopians," *BBC News*, March 26, 2009, http://news.bbc.co.uk/2/hi/africa/7959444.stm.

16
Sun, Silt, and Stagnant Ponds

1. Mesfin Mekonnen and Arjen Hoekstra, *The Water Footprint of Electricity from Hy- dropower*, Value of Water Research Report No. 51 (Delft, Netherlands: UNESCO- IHE, 2011), http://waterfootprint.org/media/downloads/Report51-WaterFootprint Hydropower_1.pdf.

2. Igor Shiklomanov, *World Water Resources: A New Appraisal and Assessment for the 21st Century* (Paris: UNESCO, 1998), http://www.ce.utexas.edu/prof/mckinney /ce385d/Papers/Shiklomanov.pdf.

3. "India's NHPC Reveals 600 MW Floating PV Plan at Hydro Dam in Maharashtra," *PV Magazine*, June 27, 2016, https://www.pv-magazine.com/2016/06/27/indias-nhpc -reveals-600-mw-floating-pv-plan-at-hydro-dam-in-maharashtra_100025162.

4. Philip Fearnside, "Hydroelectric Dams in the Brazilian Amazon as Sources of 'Greenhouse' Gases," *Environmental Conservation* 22 (1995): 7–19.
5. Luiz Pinguelli Rosa et al., "Greenhouse Gas Emissions from Hydroelectric Reservoirs in Tropical Regions," *Climate Change* 66 (2004): 9–21.
6. World Commission on Dams, *Dams and Development: A New Framework for Decision-Making* (London: Earthscan, 2000), available via UN Environment Programme, http://staging.unep.org/dams/WCD/report/WCD_DAMS%20report.pdf.
7. Vincent St. Louis et al., "Reservoir Surfaces as Sources of Greenhouse Gases to the Atmosphere," *Bioscience* 50 (2000): 766–75.
8. John Harrison et al., "Greenhouse Gas Emissions from Reservoir Water Surfaces: A New Global Synthesis," *Bioscience* 66 (2016): 949–64.
9. Fred Pearce, "Hydroelectric Dams Stoke Global Warming," *New Scientist*, June 3, 2000, https://www.eurekalert.org/pub_releases/2000-05/NS-Hdsg-3005100.php.
10. K. Mahmood, "Reservoir Sedimentation: Impact, Extent, and Mitigation," World Bank Technical Paper no. 71, September 1987, http://documents.worldbank.org/curated/en/888541468762328736/pdf/multi-page.pdf.
11. *Sedimentation and Sustainable Use of Reservoirs and River Systems* (Paris: ICOLD, 2009), http://www.icold-cigb.org/userfiles/files/CIRCULAR/CL1793Annex.pdf.
12. Taylor Maavara et al., "Global Perturbation of Organic Carbon Cycling by River Damming," *Nature Communications*, 2017, https://www.nature.com/articles/ncomms15347.

17

China

1. Gong Jing and Cui Zheng, "China's Thirst for Water Transfer," Chinadialogue, January 10, 2012, https://www.chinadialogue.net/article/show/single/en/4722.
2. J. Y. Chen et al., "Use of Water Balance Calculation and Tritium to Examine the Dropdown of Groundwater Table in the Piedmont of the North China Plain (NCP)," *Environmental Geology* 44 (2003): 564–71.
3. Junsheng Nie et al., "Loess Plateau Storage of Northeastern Tibetan Plateau-Derived Yellow River Sediment," *Nature Communication* (2015), https://www.nature.com/articles/ncomms9511.
4. *Reservoir Sedimentation*, Report no. SR 62 (Wallingford, UK: Hydraulics Research, 1985), http://eprints.hrwallingford.co.uk/1126/1/SR62.pdf.
5. Yafeng Wang et al., "Check Dam in the Loess Plateau of China," *Environmental Science & Technology* 45, no. 24 (2011), dx.doi.org/10.1021/es2038992.
6. Philip Ball, *The Water Kingdom: A Secret History of China* (London: Bodley Head, 2016).
7. Mark Giordano et al., *Water Management in the Yellow River Basin: Background, Current Critical Issues, and Future Research Needs* (Colombo, Sri Lanka: IMWI, 2004), http://www.iwmi.cgiar.org/assessment/files/pdf/publications/Research Reports/CARR3.pdf.
8. Liu Xiaoyan et al., "Healthy Yellow River's Essence and Indicators," *Journal of Geographical Sciences* 16 (2006): 259.
9. Jean Chua, "Keeping the Yellow River Flowing," *Eco-Business*, October 6, 2015, http://www.eco-business.com/news/keeping-the-yellow-river-flowing.
10. Ball, *The Water Kingdom*.

18
Unleashing the River Dragon

1. Dai Qing, *The River Dragon Has Come!: The Three Gorges Dam and the Fate of China's Yangtze River and Its People*, ed. John G. Thibodeau and Philip B. Williams (New York: M. E. Sharpe, 1998).

2. Neelam Jain, "World Bank Report Condemns India's Dams," UPI, April 6, 1995, http://www.upi.com/Archives/1995/04/06/World-Bank-report-condemns-Indias -dams/8337797140800.

3. Utpal Sandesara and Tom Wooten, *No One Had a Tongue to Speak: The Untold Story of One of History's Deadliest Floods* (Amherst, NY: Prometheus Books, 2011).

4. Richard Mahapatra, "Why Hirakud Dam Failed to Check Flood," *Down to Earth*, September 14, 2011, http://www.downtoearth.org.in/news/why-hirakud-dam-failed -to-check-flood-33982.

5. Fred Pearce, "Dams and Floods," WWF, 2001, http://reliefweb.int/report/bangladesh /dams-accused-role-flooding-research-paper-dams-and-floods.

6. "Saudi Rescue for Syria Flood Village," *BBC News,* June 17, 2002, http://news.bbc .co.uk/1/hi/world/middle_east/2050415.stm.

7. "French Firm Tasked to Repair Kariba Wall," *New Zimbabwe*, February 13, 2017, http://www.newzimbabwe.com/business-34710-French+firm+tasked+to+repair +Kariba+Dam+wall/business.aspx.

8. Fan Xiao, *Did the Zipingpu Dam Trigger China's 2008 Earthquake?: The Scientific Case* (Toronto: Probe International, 2012), http://probeinternational.org/library /wp-content/uploads/2012/12/Fan-Xiao12-12.pdf; Shemin Ge et al., "Did the Zip-ingpu Reservoir Trigger the 2008 Wenchuan Earthquake?" *Geophysical Research Letters* 36 (2009), http://onlinelibrary.wiley.com/doi/10.1029/2009GL040349 /abstract; Sharon LaFraniere, "Possible Link Between Dam and China Quake," *New York Times*, February 5, 2009, http://www.nytimes.com/2009/02/06/world /asia/06quake.html.

9. "Troops Sent to Protect China Dam," *BBC News,* September 14, 2004, http://news .bbc.co.uk/1/hi/world/asia-pacific/3654772.stm.

19
Changing Climate

1. Fiona Harvey, "Nine-Tenths of England's Floodplains Not Fit for Purpose, Study Finds," *Guardian*, June 1, 2017, https://www.theguardian.com/environment/2017 /jun/01/englands-90-floodplains-not-fit-for-purpose-study-finds.

2. Lucy Barker, "Briefing Note: Severity of the December 2015 Floods," Center for Ecology and Hydrology, January 15, 2016, https://www.ceh.ac.uk/news-and-media /blogs/briefing-note-severity-december-2015-floods-preliminary-analysis.

3. "2015 December Extreme Weather in the UK," Climateprediction.net, 2016.

4. Yukiko Hirabayashi et al., "Global Flood Risk Under Climate Change," *Nature Climate Change* 3 (2013): 816–21.

5. Colin Kelley et al., "Climate Change in the Fertile Crescent and Implications of the Recent Syrian Drought," *PNAS* 112 (2015): 3241–46.

6. Nadhir Al-Ansari et al., "Management of Water Resources in Iraq: Perspectives and Prognoses," *Engineering* 5 (2013): 667–84.

7. Fred Pearce, "Fertile Crescent Will Disappear This Century," *New Scientist*, July 29, 2009, https://www.newscientist.com/article/mg20327194.200-fertile-crescent-will -disappear-this-century.

8. "Climate Change Predicted to Increase Nile Flow Variability," *Science Daily*, April 24, 2017, https://www.sciencedaily.com/releases/2017/04/170424141236.htm.

9. *The Regional Impacts of Climate Change* (Geneva: IPCC, 1997), http://www.ipcc.ch /ipccreports/sres/regional/index.php?idp=18.

10. Michelle van Vliet et al., "Power-Generation System Vulnerability and Adaptation to Changes in Climate and Water Resources," *Nature Climate Change* 6 (2016): 375–80.

11. Jessica Belt, "How Will Latin America Deal with Its Hydropower Problem?" *Greenbiz*, May 20, 2015, https://www.greenbiz.com/article/how-will-latin-america -deal-its-hydropower-problem.

12. Daniel Farinotti et al., "From Dwindling Ice to Headwater Lakes: Could Dams Replace Glaciers in the European Alps?" *Environmental Research Letters* 11 (2016), http://iopscience.iop.org/article/10.1088/1748-9326/11/5/054022.

13. Fred Pearce, "Western Canada's Glaciers May All but Vanish by 2100," *New Scientist*, April 6, 2015, https://www.newscientist.com/article/dn27296.

14. Hamish Pritchard, "Asia's Glaciers Are a Regionally Important Buffer Against Drought," *Nature* 545 (2017): 169–74.

20
Honduras

1. "Berta Caceres," Goldman Environment Prize, http://www.goldmanprize.org /recipient/berta-caceres.

2. "Honduras: The Deadliest Country in the World for Environmental Activism," Global Witness, 2017, https://www.globalwitness.org/en/campaigns/environmental -activists/honduras-deadliest-country-world-environmental-activism.

3. "Agua Zarca Hydro Project," Banktrack, https://www.banktrack.org/project/agua _zarca_dam.

4. "River Defender Murdered in Guatemala," International Rivers, January 18, 2017, https://www.internationalrivers.org/blogs/433/river-defender-murdered-in-guatemala.

5. COPINH, https://www.copinh.org.

21
Palestine

1. Mark Zeitoun, *Power and Water in the Middle East: The Hidden Politics of the Palestinian-Israeli Water Conflict* (London: I. B. Tauris, 2008).

2. "Memorandum Submitted by Oxfam," UK Parliament Select Committee on International Development, 2004, https://www.publications.parliament.uk/pa/cm200304 /cmselect/cmintdev/230/230we34.htm.

3. Isabel Kerschner, "Radical Settlers Take on Israel," *New York Times*, September 25, 2008, http://www.nytimes.com/2008/09/26/world/middleeast/26settlers.html; Jacob Magid, "Yitzhar Settler Ordered Away from West Bank for 4 Months," *Times of Israel*, May 17, 2017, http://www.timesofisrael.com/yitzhar-settler-ordered-away -from-west-bank-for-4-months.

4. Amir Hass, "Vandals Foul Water Supplies of Palestinian Village Near Yitzhar," *Haaretz*, February 21, 2005, http://www.haaretz.com/vandals-foul-water-supply-of-palestinian-village-near-yitzhar-1.150849.

5. Amira Hass, "Israel Blocking Plan to Double Water Supply to West Bank," *Haaretz*, July 9, 2016, http://www.haaretz.com/israel-news/.premium-1.729777.

6. Haim Gvirtzman, "The Truth Behind the Palestinian Water Libels," BESA, February 24, 2014, https://besacenter.org/perspectives-papers/truth-behind-palestinian-water-libels.

7. "West Bank Barrier Threatens Villagers' Way of Life," *BBC News,* May 10, 2012, http://www.bbc.co.uk/news/magazine-18012895.

22
River Jordan

1. Greg Shapland, *Rivers of Discord: International Water Disputes in the Middle East* (London: Hurst, 1997).

2. Munther Haddadin, "Water in the Middle East Peace Process," *Geographical Journal* 168 (2002): 324–40.

3. Ariel Sharon, *Warrior: The Autobiography of Ariel Sharon* (London: Simon & Schuster, 2001).

4. Martin Sherman, *The Politics of Water in the Middle East: An Israeli Perspective on the Hydro-Political Aspects of the Conflict* (New York: St. Martin's Press, 1999).

5. Noa Shpigel and Zafrir Rinat, "Syrian Civil War Has Several Environmental Side-Effects on Israel and Jordan," *Haaretz*, December 12, 2016, http://www.haaretz.com/middle-east-news/jordan/.premium-1.758168.

6. Zafrir Rinat, "Israel's National Water Carrier: Both Boom and Bane," *Haaretz*, June 26, 2014, http://www.haaretz.com/israel-news/science/.premium-1.601284.

7. Rowan Jacobsen, "How a New Source of Water Is Helping Reduce Conflict in the Middle East," *Ensia*, July 19, 2016, https://ensia.com/features/water-desalination-middle-east.
   David Brooks et al., "Changing the Nature of Transboundary Water Agreements: The Israeli–Palestinian Case," *Water International* 38 (2013), http://www.tandfonline.com/doi/abs/10.1080/02508060.2013.810038; *Take Me Over the Jordan,* EcoPeace/FoEME, 2012, http://foeme.org/uploads/13480625611˜%5E$%5E˜Take_Me_Over_the_Jordan_2012_WEB.pdf.

8. Arie Issar and Eilon Adar, "Progressive Development of Water Resources in the Middle East for Sustainable Water Supply in a Period of Climate Change," *Philosophical Transactions of the Royal Society A* 368 (2010): 5339–50.

23
Egypt

1. Robert O. Collins, *The Nile* (New Haven, CT: Yale University Press, 2002); Paul Howell and Tony Allan, *The Nile: Sharing a Scarce Resource* (Cambridge, UK: Cambridge University Press, 1994).

2. "Egypt Demands Ethiopia Halt Nile Dam, Upping Stakes," Reuters, June 5, 2013, http://www.reuters.com/article/us-egypt-ethiopia-dam-idUSBRE9541EQ20130605.

3. Dale Whittington, John Waterbury, and Marc Jeuland, "The Grand Renaissance Dam and Prospects for Cooperation on the Eastern Nile," paper, 2014, available

from Duke Global Health Institute Website Network, https://sites.globalhealth .duke.edu/jeulandresearch/wp-content/uploads/sites/5/2015/07/Whittington -Waterbury-Jeuland-Feb-10-2014-Grand-Renaissance-Dam.pdf.

4. "Egypt Stresses Nile Water Rights in Ethiopia Dam Project," AFP, March 25, 2015, https://www.yahoo.com/news/egypt-stresses-nile-water-rights-ethiopia-dam-project -191717687.html.

5. Mengisteab Teshome, "Ethiopia: GERD Lottery with 25 Million Birr to Be on Sale," *Ethiopian Herald*, April 6, 2017, http://allafrica.com/stories/201704060699.html.

6. Abdul Latif Jameel/World Water and Food Security Lab, *The Grand Ethiopian Renaissance Dam: An Opportunity for Collaboration and Shared Benefits in the Eastern Nile Basin* (Cambridge, MA: MIT, 2014), http://jwafs.mit.edu/sites/default/files /documents/GERD_2014_Full_Report.pdf.

7. "Cooperative Framework Agreement," Nile Basin Initiative, http://www.nilebasin .org/index.php/nbi/cooperative-framework-agreement.

24
Iraq

1. Danya Chudacoff, "'Water War' Threatens Syria Lifeline," Al Jazeera, July 7, 2014, http://www.aljazeera.com/news/middleeast/2014/07/water-war-syria-euphrates -2014757640320663.html.

2. "IS Conflict: Raqqa Warning over 'Risk to Tabqa Dam,'" *BBC News,* March 26, 2017, http://www.bbc.co.uk/news/world-middle-east-39399803; Tom Miles, "UN Warns of Catastrophic Dam Failure in Syria Battle," Reuters, February 15, 2017, http://uk.reuters.com/article/uk-mideast-crisis-syria-dam-idUKKBN15U1CC.

3. Ben Smith, "Worsening Humanitarian Crisis in Syria and Iraq," House of Commons Library, July 8, 2014, http://researchbriefings.parliament.uk/ResearchBriefing /Summary/SN06926.

4. Alissa Rubin and Rod Nordland, "Sunni Militants Advance Toward Large Iraqi Dam," *New York Times,* June 25, 2014, https://www.nytimes.com/2014/06/26/world /middleeast/isis-iraq.html.

5. Dexter Filkins, "A Bigger Problem Than ISIS?" *New Yorker,* January 2, 2017, http:// www.newyorker.com/magazine/2017/01/02/a-bigger-problem-than-isis.

6. "Iraq Awards Italian Firm Trevi Contract to Repair Mosul Dam," AFP, February 2, 2016, https://www.theguardian.com/world/2016/feb/02/iraq-awards-italian-firm -trevi-contract-repair-mosul-dam.

7. Rikar Hussain, "Mosul Dam No Longer on Brink of Catastrophe," Voice of America, May 5, 2017, http://www.voanews.com/a/mosul-dam-no-longer-brink -catastrophe/3839850.html.

8. Jim Michaels, "Mosul Dam Risks Devastating Failure as Iraq Government Keeps Stalling," *USA Today,* May 30, 2017, https://www.usatoday.com/story/news/world /2017/05/30/iraq-government-has-yet-decide-if-renew-critical-work/102198756.

25
Mesopotamia

1. Wilfred Thesiger, *The Marsh Arabs* (Harmondsworth, UK: Penguin, 1964).

2. House of Commons Debate, January 21, 1993, http://hansard.millbanksystems.com /commons/1993/jan/21/the-gulf-and-raf.

3. Hassan Partow, *The Mesopotamian Marshes: Demise of an Ecosystem* (Geneva: UNEP, 2001), http://www.grid.unep.ch/activities/sustainable/tigris/mesopotamia .pdf.

4. Fred Pearce, "We Can Save Iraq's Garden of Eden," *New Scientist*, April 17, 2013, https://www.newscientist.com/article/mg21829130-200.

5. "Iraq, ISIS Angered as Turkey Hoards Euphrates Waters," *Daily Star* (Lebanon), July 6, 2015, http://www.dailystar.com.lb/News/Middle-East/2015/Jul-06/305304 -iraq-isis-angered-as-turkey-hoards-euphrates-waters.ashx.

6. "ISIS 'Causing Drought' by Cutting Water Supplies in Four Iraqi Provinces," *Middle East Monitor*, June 19, 2015, https://www.middleeastmonitor.com/20150619 -isis-causing-drought-by-cutting-water-supplies-in-four-iraqi-provinces.

7. Kamal Chomani and Toon Bijnens, *The Impact of the Daryan Dam on the Kurdistan Region of Iraq* (Save the Tigris and Iraqi Marshes Campaign, October 2016), http://www.iraqicivilsociety.org/wp-content/uploads/2016/10/Daryan-Dam -Report.pdf.

8. "University of Salford Researchers Call for Iran-Iraq Water Treaty," University of Salford, July 7, 2014, http://www.salford.ac.uk/news/articles/2015/university-of -salford-researchers-call-for-iran-iraq-water-treaty.

26
Tibet

1. "China Dams the Brahmaputra," *Rediff News* (India), October 21, 2015, http:// www.rediff.com/news/column/china-dams-the-brahmaputra-why-india-should -worry/20151021.htm.

2. Fred Pearce, "China Is Taking Control of Asia's Water Tower," *New Scientist*, April 26, 2012, https://www.newscientist.com/article/mg21428624.400.

3. "Public Statement on the Mong Ton Dam on the Salween River," International Rivers, June 10, 2016, https://www.internationalrivers.org/resources/public-statement-on-the-mong-ton-dam-on-the-salween-river-11496; Julian Kirchherr et al., "The Interplay of Activists and Dam Developers: The Case of Myanmar's Mega-Dams," *International Journal of Water Resources Development* (2016), http:// www.geog.ox.ac.uk/graduate/research/jkirchherr-150410.pdf.

4. "7,100 MW Bunji Dam's Detailed Engineering Design Completed," *Paktribune*, May 7, 2013, http://paktribune.com/business/news/7100-MW-Bunji-Dams-detailed -engineering-design-completed-11159.html.

5. Yong Jhong et al., "Rivers and Reciprocity: Perceptions and Policy on International Watercourses," *Water Policy* (2016), http://wp.iwaponline.com/content/early /2016/02/29/wp.2016.229.

6. Benjamin Pohl, *The Rise of Hydro-Diplomacy: Strengthening Foreign Policy for Transboundary Waters* (Berlin: Climate Diplomacy, 2014), https://www.adelphi.de /en/system/files/mediathek/bilder/the_rise_of_hydro-diplomacy_adelphi.pdf.

7. "India Rejects Pak's Claims of Halting Miyar Dam Project," *Rediff News*, March 22, 2017, http://www.rediff.com/news/report/india-rejects-paks-claims-of-halting -miyar-dam-project/20170322.htm.

8. "Water Conflicts Between Asian Nuclear Powers Pose Global Threat," Phys.org, October 28, 2016, https://phys.org/news/2016-10-conflicts-asian-nuclear-powers -pose.html.

9. Indrani Bagchil, "India Refuses to Change Miyar Dam Design, *Times of India,* March 24, 2017, http://timesofindia.indiatimes.com/india/india-refuses-to-change -miyar-dam-design-will-stick-to-letter-of-the-indus-waters-treaty/articleshow /57815005.cms; Mubarak Zeb Khan, "India Asked to Stop Work on Kishanganga and Ratle Projects," *Dawn,* January 21, 2017, https://www.dawn.com/news/1309767.

27
Elisha's Spring and the Riddle of Angkor

1. "Neolithic Jericho and the Origins of Civilization," Auja Ecocenter, http://www .aujaecocenter.org/index.php/the-story-of-the-valley/neolithic-jericho-and-the -origins-of-civilization, accessed 2017.
2. Peter Gleick, "Water, War and Peace in the Middle East," *Environment: Science and Policy for Sustainable Development* 36 (1994): 6–42.
3. Leonard Woolley, *Ur of the Chaldees* (London: Ernest Benn, 1929).
4. Edward Goldsmith and Nicholas Hildyard, "Traditional Irrigation in Meso- potamia," http://archive.li/nbwOy, originally published in their *The Social and Environmental Effects of Large Dams,* vol. 1 (Cornwall, UK: Wadebridge Ecological Centre, 1984).
5. Thorkild Jacobsen and Robert Adams, "Salt and Silt in Ancient Mesopotamian Agriculture," *Science* 128 (1958): 1251–58.
6. William Willcocks, *Ancient System of Irrigation in Bengal* (Delhi: B. R. Publishing, 1930).
7. Karl Wittfogel, *Oriental Despotism* (New Haven, CT: Yale University Press, 1957).
8. Damian Evans et al., "A Comprehensive Archaeological Map of the World's Larg- est Preindustrial Settlement Complex at Angkor, Cambodia," *PNAS* 104 (2007), doi: 10.1073/pnas.0702525104.
9. Matti Kummu, "Water Management in Angkor," *Journal of Environmental Management* 90 (2009); 1413–21.
10. Matti Kummu, "The Natural Environment and Historical Water Management of Angkor, Cambodia," paper, 2003, http://citeseerx.ist.psu.edu/viewdoc/download ?doi=10.1.1.452.9893&rep=rep1&type=pdf.
11. Lisa Lucero, "The Collapse of the Classic Maya: A Case for the Role of Water Con- trol," *American Anthropologist* 104 (2002): 814–26; Jared Diamond, *Collapse: How Societies Choose to Fail or Succeed* (New York: Viking, 2005).

28
Aral Sea

1. Grigori Reznichenko, *The Aral Sea Tragedy: The Diary of an Expedition* (Moscow: Novosti, 1992).
2. Geoffrey Lean, "The Dead Sea That Sprang to Life," *Independent,* May 28, 2006, http://www.schwartzreport.net/the-dead-sea-that-sprang-to-life.
3. Enjoli Liston, "Satellite Images Show Aral Sea Basin 'Completely Dried,'" *Guard- ian,* October 1, 2014, https://www.theguardian.com/world/2014/oct/01/satellite -images-show-aral-sea-basin-completely-dried.
4. Philip Micklin, "Desiccation of the Aral Sea: A Water Management Disaster in the Soviet Union," *Science* 241 (1988): 1170–76.

5. Matt Spetalnick et al., "U.S. Lifted Uzbekistan's Rights Ranking as Cotton Field Abuses Continued," Reuters, December 23, 2015, http://uk.reuters.com/article/us -usa-humantrafficking-uzbekistan-insig-idUKKBN0U60EZ20151223; "Uzbekistan's Forced Labor Problem," International Labor Rights Forum, 2017, http://www .cottoncampaign.org/uzbekistans-forced-labor-problem.html.

6. Olli Varis, "Resources: Curb Vast Water Use in Central Asia," *Nature*, October 1, 2014, http://www.nature.com/news/resources-curb-vast-water-use-in-central -asia-1.16017#/b1.

7. Sarah O'Hara et al., "Exposure to Airborne Dust Contaminated with Pesticide in the Aral Sea," *Lancet* (February 19, 2000), http://dx.doi.org/10.1016/S0140-6736 (99)04753-4; *Karakalpakstan: A Population in Danger* (Uzbekistan: Medecins Sans Frontieres, 2003–2004), https://www.aerzte-ohne-grenzen.de/sites/germany/files /attachments/2003-04-karakalpakstan-report-population-in-danger.pdf.

8. Oral Ataniyazova, "Health and Ecological Consequences of the Aral Sea Crisis," paper prepared for 3rd World Water Forum, Regional Cooperation in Shared Water Resources in Central Asia, Kyoto, March 18, 2003, http://www.caee.utexas .edu/prof/mckinney/ce385d/papers/atanizaova_wwf3.pdf; Oral Ataniyazova, "Sea of Troubles," *Index on Censorship* 27 (1998): 142–46.

9. Brett Walton, "Natural Gas Found in Aral Sea's Dry Bed," Circle of Blue, June 11, 2010, http://www.circleofblue.org/2010/asia/natural-gas-found-in-aral-sea%E2%80 %99s-dry-bed.

10. Aynur Jafarova, "Turkmenistan Creates Man-Made Lake in Karakum Desert," *AzerNews*, August 27, 2013, https://www.azernews.az/region/58638.html.

11. "Turkmenistan to Create Desert Sea," *BBC News,* July 16, 2009, http://news.bbc .co.uk/1/hi/world/asia-pacific/8154467.stm.

29

Harvesting the Monsoons

1. Zhu Qiang and Li Yuanhong, "Rainwater Harvesting in the Loess Plateau of Gansu, China, and Its Significance," Gansu Research Institute for Water Conservancy, 2003, http://www.eng.warwick.ac.uk/ircsa/pdf/9th/02_03.pdf.

2. N. S. Vangani, K. D. Sharma, and P. C. Chatterji, *Tanka—A Reliable System of Rainwater Harvesting in the Indian Desert* (Jodhpur: CAZRI, 1988), http://krishikosh .egranth.ac.in/bitstream/1/02050147/1/60-(TANKA-A%20RELIABLE%20SYSTEM %20OF%20RAINWATER%20).pdf.

3. Anil Agarwal, *Dying Wisdom: The Decline and Revival of Traditional Water Harvesting Systems in India* (Delhi: Centre for Science and the Environment, 1997).

4. Rajendra Singh, "The River Maker," *New Scientist*, September 7, 2002, https:// www.newscientist.com/article/mg17523595-600.

5. "The Waterman of India," Common Threads, 2015, http://commonthreads.sgi.org /post/150381747173/the-waterman-of-indiareviving-arid-lands-with.

6. Roger Harrabin, "'Water Man of India' Rajendra Singh Bags Top Prize," *BBC News,* March 21, 2015, http://www.bbc.co.uk/news/science-environment-32002306.

7. Meera Subramanian, *A River Runs Again: India's Natural World in Crisis* (New York: Public Affairs, 2015).

8. Theib Oweis, *Water Harvesting: Indigenous Knowledge for the Future of the Drier Environments* (Aleppo: ICARDA, 2001).

9. Theib Oweis, *Indigenous Water-Harvesting Systems in West Asia and North Africa* (Aleppo: ICARDA, 2004).

10. "Examples of Rainwater Harvesting and Utilisation Around the World," UNEP, http://www.unep.or.jp/ietc/publications/urban/urbanenv-2/9.asp.

11. Anil Agarwal, "Coaxing the Barren Deserts Back to Life," *New Scientist*, September 15, 1977.

12. Fred Pearce, "Interview: Can't See the Desert for the Trees," *New Scientist*, March 26, 2008, https://www.newscientist.com/article/mg19726491-700.

13. Mary Tiffen, Michael Mortimore, and Francis Gichuki, *More People, Less Erosion: Environmental Recovery in Kenya* (London: Overseas Development Institute, 1994).

14. Gorrel Espelund, "Successful Green Revolution in Africa," *Waterfront*, SIWI, May 2017, http://www.siwi.org/publications/stockholm-water-front-no-1-2017.

15. Dieter Prinz, "Water Harvesting Past and Future," in *Sustainability of Irrigated Agriculture*, ed. L. S. Pereira et al., NATO Science Series (Springer Netherlands, 1996).

30
Oman

1. Lynn Teo Simarski, "Oman's Unfailing Springs," *Aramco World*, 1992, http://archive.aramcoworld.com/issue/199206/oman.s.unfailing.springs.htm.

2. Slim Zekri and Ahmed Salim Al-Marshudi, "A Millenarian Water Rights System and Water Markets in Oman," *Water International* 33 (2008), http://www.tandfonline.com/doi/abs/10.1080/02508060802256120.

3. Peter Beaumont and Michael Bonine, *Qanat, Kariz & Khattara* (London: SOAS, 1989); Dale Lightfoot, "The Origin and Diffusion of Qanats in Arabia," *Geographical Journal* 166 (2000): 215–26.

4. Harriet Nash, *Water Management: The Use of Stars in Oman* (Oxford, UK: British Archaeological Reports, 2011).

5. "Omani Falajs [*sic*] in World Heritages List," Ministry of Regional Municipalities and Water Resources, http://mrmwr.gov.om/new/en/Page.aspx?id=10&li=-1&Type=W_Sec&Slide=false&color1=03AADE&color2=7EB204.

6. "Ministry Embarks on Huge Project to Maintain the Aflaj," *Times of Oman*, May 5, 2014, http://timesofoman.com/article/33942/Oman/Ministry-embarks-on-huge-project-to-maintain-the-aflaj.

7. Anthony Smith, *Blind White Fish in Persia* (London: Allen & Unwin, 1953).

31
Iran

1. *Qanat Irrigation Systems* (Tehran, Iran: Cenesta, 2003), ftp://ftp.fao.org/sd/SDA/GIAHS/final_qanats_proposal.pdf.

2. Paul Ward English, "Qanats and Lifeworlds in Iranian Plateau Villages," *Yale F&ES Bulletin* 103, https://environment.yale.edu/publication-series/documents/downloads/0-9/103english.pdf, accessed 2017.

3. Dale Lightfoot, *Survey of Infiltration Karez in Northern Iraq* (Paris: UNESCO, 2009), http://unesdoc.unesco.org/images/0018/001850/185057E.pdf.

4. Frederic Pellet et al., "Geotechnical Performance of Qanats during the 2003 Bam, Iran, Earthquake," *Earthquake Spectra* 21 (2005): S137–64.

5. "Lost Kingdom of the Sahara," *BBC News,* July 21, 2000, http://news.bbc.co.uk/1/hi/world/middle_east/845160.stm.

6. Lightfoot, *Survey of Infiltration Karez in Northern Iraq.*

7. J. Wessels and R. J. A. Hoogeveen, "Renovation of Qanats in Syria," paper presented at UNU/UNESCO/ICARDA international workshop, Alexandria, Egypt, September 2002, http://www.waterhistory.org/histories/qanatrenovation/renovation.pdf; "Land Evaluation: Towards a Revised Framework," Land and Water Discussion Paper 6, FAO, 2007, http://www.fao.org/3/a-a1080e.pdf.

8. Dale Lightfoot, "Syrian Qanat Romani: History, Ecology, Abandonment," *Journal of Arid Environments* 33 (1996): 321–36.

9. Dale Lightfoot, "Syrian Qanat Romani," Water History, n.d., http://waterhistory.org/histories/syria/syrian.pdf.

10. Zvi Ron, "Development and Management of Irrigation Systems in Mountain Regions of the Holy Land," *Transactions of the Institute of British Geographers* 10 (1985): 149–69.

11. "Karez Wells," UNESCO, 2008, http://whc.unesco.org/en/tentativelists/5347.

12. Daanish Mustafa, "The Necessity of Karez Water Systems in Balochistan," Middle East Institute, 2014, http://www.mei.edu/content/necessity-karez-water-systems-balochistan.

13. John Mervin, "In Peru, Water Is a High Price for Lima's Poor," *BBC News,* October 9, 2015, http://www.bbc.co.uk/news/business-34451418.

14. "Lima to Restore Pre-Incan Water Management System to Confront Water Crisis," Forest Trends, press release, April 8, 2015, http://www.forest-trends.org/documents/files/doc_4893.pdf.

15. Gena Gammie and Bert de Bievre, *Assessing Green Interventions for the Water Supply of Lima, Peru: Cost-Effectiveness, Potential Impact, and Priority Research Areas,* Forest Trends, 2015, http://www.forest-trends.org/documents/files/doc_4896.pdf.

32
Taking the Water to the People

1. Jonathan Kaiman, "China's Water Diversion Project Starts to Flow to Beijing," *Guardian,* December 12, 2014, https://www.theguardian.com/world/2014/dec/12/china-water-diversion-project-beijing-displaced-farmers.

2. T. V. Padma, "India's Grand Plan to Create World's Longest River Set to Go," *New Scientist,* November 28, 2016, https://www.newscientist.com/article/2114431-indias-grand-plan-to-create-worlds-longest-river-set-to-go.

3. Manu Balachandran, "Why India's $168 Billion River-Linking Project Is a Disaster-in-Waiting," *Quartz,* September 18, 2015, https://qz.com/504127/why-indias-168-billion-river-linking-project-is-a-disaster-in-waiting.

4. Sachin Saini, "Rajasthan Set to Begin Work on Its First River-Interlinking Project," *Hindustan Times,* May 1, 2017, http://www.hindustantimes.com/jaipur/rajasthan-set-to-begin-work-on-its-first-river-linking-project/story-pwqdm7ABBqdxingcEyhNvI.html.

5. Navin Singh Khadka, "India to Divert Rivers to Tackle Drought," *BBC News,* May 16, 2016, http://www.bbc.co.uk/news/world-asia-india-36299778.

6. H. S. Mangat, "Water War Between Junjab and Haryana," *Economic & Political Weekly,* December 10, 2016, http://www.epw.in/journal/2016/50/special-articles/water-war-between-punjab-and-haryana.html.

7. "Why Water War Has Broken Out in India's Silicon Valley," *BBC News*, September 13, 2016, http://www.bbc.co.uk/news/world-asia-india-37346570.

8. Pearce, "China Is Taking Control of Asia's Water Tower."

9. Thomas Erdbrink, "The Empty River of Life," *New York Times*, May 5, 2015, https://www.nytimes.com/2015/05/06/world/middleeast/iran-our-man-in-tehran.html.

10. Abah Adah, "Chad: Nigeria, Chad Seek $50 Billion Aid to Recharge Shrinking Lake Chad," *All Africa*, January 30, 2017, http://allafrica.com/stories/201701300189.html.

11. *Joint Environmental Audit on the Drying Up of Lake Chad.*

33
Singapore

1. "Singapore-Malaysia Water Agreements," National Library Board Singapore, updated 2011, http://eresources.nlb.gov.sg/infopedia/articles/SIP_1533_2009-06-23.html.

2. "India Water Supply and Sanitation," World Bank, 2006, http://documents.worldbank.org/curated/en/150271468756313924/pdf/358360REVISED0IN0Urban01PUBIC1.pdf.

3. Linus Dagerskog, "Productive Sanitation: Reuse as a Driver to Take Sanitation to Scale in the Sahel," presentation from the 2012 World Water Week in Stockholm, http://www.worldwaterweek.org/documents/WWW_PDF/2012/Thur/No-Food-and-Nutrition/Linus-Dagerskog-Moussa-Bonzi.pdf.

4. Fred Pearce, "Flushed with Success," *New Scientist*, February 16, 2013, https://www.newscientist.com/article/mg21729042-200.

5. Christopher Scott, "Confronting the Realities of Wastewater Use in Agriculture," IWMI, *Water Policy Briefing* 9, 2003, http://www.iwmi.cgiar.org/Publications/Water_Policy_Briefs/PDF/wpb09.pdf.

6. Fred Pearce, "Sewage Waters a Tenth of World's Irrigated Crops," *New Scientist*, August 18, 2004, https://www.newscientist.com/article/dn6297.

7. Andy Coghlan, "Faeces of Food Crops Safer Than You'd Think," *New Scientist*, September 13, 2006, https://www.newscientist.com/article/mg19125694-400.

8. Jeena Srinivasan and Vatna Reddy, "Impact of Irrigation Water Quality on Human Health," *Ecological Economics* 68 (2009): 2800–2807.

9. "Guidelines on the Use of Urine and Faeces in Crop Production," EcoSanRes, 2004, http://www.ecosanres.org/pdf_files/ESR-factsheet-06.pdf.

10. Michael Gilmont et al., *Decoupling National Water Needs for National Water Supplies: Insights and Potential for Countries in the Jordan Basin* (Amman, Jordan: WANA Institute, 2017), http://wanainstitute.org/sites/default/files/publications/Publication_DeliveringFoodAndWaterSecurity_0.pdf.

34
Out of Thin Air

1. Edward Martin, "Dew Ponds," http://dewponds.co.uk/articles_dewponds_martin.htm, 2006.

2. Gilbert White, *The Natural History of Selborne* (Harmondsworth, UK: Penguin Classics, 1977).

3. "The Dewpond," Cleeve Common Board of Conservators, 2013, https://static1.squarespace.com/static/56c466381d07c0cc6a57a634/t/56f01c6486db4349d8860ed6/1458576485561/dewpond-0088.pdf.

4. "The Spirit of the Downs," http://dewponds.co.uk/biblio_spirit_of_the_downs.htm.

5. Marc Muselli et al., "Dew and Rain Water Collection in the Dalmatian Coast, Croatia," *Atmospheric Research* 92 (2009): 455–63.

6. "Volcanoes, Eruptions, Lanzarote," Lanzarote Information, 2009, http://www .lanzaroteinformation.com/content/volcanoes-eruptions-lanzarote.

7. David Riebold, *Stone Mulches and Occult Precipitation* (Bangor, UK: University of Wales, 1993).

8. Robert Nelson, "Air Wells, Fog Fences & Dew Ponds," RexResearch, 2003, http:// www.rexresearch.com/airwells/airwells.htm.

9. "Sustainable Water Solutions," Fogquest, http://www.fogquest.org/project -information/current-projects.

10. Andrew Parker and Chris Lawrence, "Water Captured by Desert Beetle," *Nature* 414 (2001), 33–34.

11. Anne Trafton, "Beetle Spawns New Material," MIT News, June 14, 2006, http:// news.mit.edu/2006/beetles-0614; Nancy Owano, "Self-Filling Water Bottle Takes Cues from Desert Beetle," Phys.org, November 25, 2012, https://phys.org/news /2012-11-self-filling-bottle-cues-beetle.html.

35
Seeding Clouds and Desalting the Sea

1. "Iran to Use Drones for Cloud Seeding," *Financial Tribune*, May 4, 2017, https:// financialtribune.com/articles/people-environment/63651/iran-to-use-drones-for -cloud-seeding.

2. "Idaho Power's Cloud Seeding Program," Idaho Power, https://www.idahopower .com/pdfs/AboutUs/EnergySources/FAQ_CloudSeeding_12-14.pdf; "UW to Lead Cloud-Seeding Project in Southwestern Idaho," University of Wyoming, January 13, 2017, http://www.uwyo.edu/uw/news/2017/01/uw-to-lead-cloud-seeding -project-in-southwestern-idaho.html.

3. Yasmin Al Heialy, "Revealed: $558,000 Spent on UAE Cloud-Seeding Operations Last Year," *Arabian Business*, April 28, 2016, http://www.arabianbusiness.com /revealed-558-000-spent-on-uae-cloud-seeding-operations-last-year-629970.html.

4. Christie Aschwanden, "We May Never Know How Well Cloud Seeding Works," *FiveThirtyEight*, December 22, 2014, https://fivethirtyeight.com/features/we-may -never-know-how-well-cloud-seeding-works.

5. John Vidal and Helen Weinstein, "RAF Rainmakers 'Caused 1952 Flood,'" *Guardian*, August 30, 2001, https://www.theguardian.com/uk/2001/aug/30/sillyseason .physicalsciences.

6. Wang Xian, "China Pours Billion into Rainmaking," *China Daily*, March 24, 2011, http://www.chinadaily.com.cn/china/2011-03/24/content_12218277.htm.

7. Jonathan Watts, "Cities Fall Out over Cloud," *Guardian*, July 15, 2004, https:// www.theguardian.com/environment/2004/jul/15/china.weather.

8. Linda Poon, "The United Arab Emirates Needs More Rain, So It's Building a Mountain," *CityLab*, May 5, 2016, https://www.citylab.com/environment/2016/05 /the-united-arab-emirates-needs-more-rain-so-its-building-a-mountain/481440.

9. Sophie Bushwick, "Get Your Iceberg Water, Here," *Scientific American*, August 10, 2011, https://blogs.scientificamerican.com/observations/get-your-iceberg-water -here530-a-glass.

10. Derek Baldwin, "Firm to Tow Icebergs from Antarctica to Fujairah," *Gulf News*, May 2, 2017, http://gulfnews.com/news/uae/tourism/firm-to-tow-icebergs-from -antarctica-to-fujairah-1.2020868.

11. David Talbot, "Megascale Desalination," *MIT Technology Review* (2015), https:// www.technologyreview.com/s/534996/megascale-desalination.

12. Amanda Covarrubias, "Santa Barbara Working to Reactivate Mothballed Desalination Plant," *Los Angeles Times*, March 3, 2015, http://www.latimes.com/local /california/la-me-santa-barbara-desal-20150303-story.html.

13. Paul Rincon, "Graphene-Based Sieve Turns Seawater into Drinking Water," *BBC News*, April 3, 2017, http://www.bbc.co.uk/news/science-environment -39482342.

14. Heather Cooley et al., *Key Issues in Seawater Desalination in California: Marine Impacts* (Oakland, CA: Pacific Institute, December 2013), http://pacinst.org/wp-content/uploads/2013/12/desal-marine-imapcts-full-report.pdf.

36
Europe

1. Wendy Jones et al., *LIFE and Europe's Rivers: Protecting and Improving Our Water Resources* (Luxembourg: Office for Official Publications of the European Communities, 2007), http://ec.europa.eu/environment/life/publications/lifepublications /lifefocus/documents/rivers.pdf.

2. "Removal of the Maisons-Rouges Dam over the River Vienne," ONEMA, May 2010, http://www.onema.fr/EN/EV/publication/rex_r1_vienne_vbatGB.pdf.

3. River Restoration Centre, http://www.therrc.co.uk/river-restoration.

4. Nicholas Barton, *The Lost Rivers of London* (1962; London: Historical Publications, 1992).

5. Robert William Steel, *River Wandle Companion and Wandle Trail Guide* (London: Culverhouse Books, 2012).

6. "Thames Water Fined £20m for Sewage Spill," *BBC News*, March 22, 2017, http:// www.bbc.co.uk/news/uk-england-39352755.

7. Fred Pearce, "The Biologist Who Broke the Berlin Wall," *New Scientist*, July 8, 2009, https://www.newscientist.com/article/mg20327161-400; Fred Pearce, "Rising Waters Drown Opposition to Slovakia's Dam," *New Scientist*, July 16, 1994, https:// www.newscientist.com/article/mg14319341-200.

8. "Albanian Court Stops Dam Project on the Vjosa," Save the Blue Heart of Europe, May 3, 2017, http://balkanrivers.net/en/news/albanian-court-stops-dam-project -vjosa.

9. "Floodplain Management: Reducing Flood Risks and Restoring Healthy Ecosystems," European Environment Agency, January 26, 2016, http://www.eea.europa.eu /highlights/floodplain-management-reducing-flood-risks.

10. Harvey, "Nine-Tenths of England's Floodplains Not Fit for Purpose, Study Finds."

11. Marc Cioc, *The Rhine: An Eco-Biography* (Seattle: University of Washington Press, 2015).

12. Federal Ministry for the Environment, Nature Conservation, Building and Nuclear Safety, "Trittin Presents Draft Flood Control Act," press statement, August 8, 2003.

37
Ganges and Mississippi

1. Fred Pearce, *Confessions of an Eco-Sinner: Tracking Down the Sources of My Stuff* (Boston: Beacon Press, 2008).

2. Shahjahan Mondal et al., "Hydro-Meteorological Trends in Southwest Coastal Bangladesh," *American Journal of Climate Change* (March 2013), doi: 10.4236 /ajcc.2013.21007.

3. John Pethick and Julian Orford, "Rapid Rise in Effective Sea Level in Southwest Bangladesh," *Global and Planetary Change* 111 (2013): 237–45.

4. "Coastal Embankment Improvement Project—Phase I (CEIP-I)," World Bank, 2013, http://projects.worldbank.org/P128276/coastal-embankment-improvement -project-phase-1ceip-1?lang=en.

5. John McPhee, *The Control of Nature* (New York: Farrar, Straus, and Giroux, 1989).

6. Jim Robbins, "Why the World's Rivers Are Losing Sediment and Why It Matters," *Yale e360*, June 20, 2017, https://e360.yale.edu/features/why-the-worlds-rivers-are -losing-sediment-and-why-it-matters.

7. Tim Hirsch, "Katrina Damage Blamed on Wetlands Loss," *BBC News*, November 1, 2005, http://news.bbc.co.uk/1/hi/world/americas/4393852.stm.

8. William Sellers, "Memoir of James Buchanan Eads. 1820–1887," read before the National Academy of Sciences, 1888, available at http://www.nasonline.org /publications/biographical-memoirs/memoir-pdfs/eads-james-b.pdf.

9. Mark Twain, *Life on the Mississippi* (New York: Harper, 1901), 207.

38
More Crop per Drop

1. Natalie Kotsios, "Murray-Darling Basin Plan: John Howard's Vision Still Controversial," *Weekly Times* (Australia), January 25, 2017, http://www.weeklytimesnow .com.au/news/national/murraydarling-basin-plan-john-howards-vision-still -controversial/news-story/5ad40b31c3d4d4c4c09276672db5799d.

2. Mike Young, *Tricks of the Trade: Learning from Australia's Water Reform Experience* (Stockholm: Waterfront, SIWI, March 2015), http://freshwater.issuelab.org /resource/tricks-of-the-trade-learning-from-australia-s-water-reform-experience .html.

3. Andrew Spence, "High Flows Improve Vital Signs of Life for River Murray," *InDaily* (Australia), January 16, 2017, http://indaily.com.au/news/local/2017/01/16 /high-flows-improve-vital-signs-of-life-for-river-murray.

4. Emma Marris, "Water: More Crop per Drop," *Nature* 452 (2008): 273–77.

5. Feng Hu, "8 Reasons to Invest in Irrigation in China," ChinaWaterRisk, August 18, 2016, http://chinawaterrisk.org/resources/analysis-reviews/8-reasons-to-invest-in -irrigation-in-china.

6. "Pepsee Micro-Irrigation System," WOCAT, 2007, http://teca.fao.org/sites/default /files/technology_files/10_PepseeMicroIrrigation_India.pdf.

7. Shilp Verma et al., "Pepsee Systems: Grassroots Innovation Under Groundwater Stress," *Water Policy* 6 (2004): 303–18.

8. "More Rice with Less Water," WWF India, http://d2ouvy59p0dg6k.cloudfront.net /downloads/wwf_rice_report_2007.pdf.

9. E. Lopez-Gunn et al., "Lost in Translation? Water Efficiency in Spanish Agriculture," *Agricultural Water Management* 108 (2012): 83–95.

10. Lisa Pfeiffer and Cynthia Lin, "Does Efficient Irrigation Technology Lead to Reduced Groundwater Extraction?: Empirical Evidence," *Journal of Environmental Economics and Management* 67 (2014): 189–208.

11. Bruce Lankford, "The Paracommons of Salvaged Water," *Land* (2014), http://www .thelandmagazine.org.uk/articles/paracommons-salvaged-water; Bruce Lankford, "Fictions, Fractions, Factorials and Fractures; On the Framing of Irrigation Efficiency," *Agricultural Water Management* (2012), http://agris.fao.org/agris-search /search.do?recordID=US201400172037.

39
Colombia

1. Autumn Spanne, "Colombia's Cloud Forests Imperiled by Climate Change, Development," *Daily Climate*, December 4, 2012, available via *Scientific American*, https://www.scientificamerican.com/article/colombias-cloud-forests-imperiled -by-climate-change-development.

2. Misty Herrin, "Eight Million Residents of Bogota Obtain Water from Chingaza and Sumapaz National Parks," Nature Conservancy, https://www.nature.org /ourinitiatives/regions/southamerica/colombia/water-fund-bogota.xml, accessed May 2017.

3. "Colombia Minister in Battle over Cajamarca Mining Ban," *BBC News*, March 28, 2017, http://www.bbc.co.uk/news/world-latin-america-39425592.

4. *Water Funds: Conserving Green Infrastructure; a Guide for Design, Creation, and Operation* (Bogota, Colombia: Nature Conservancy, 2012), https://www.nature.org /media/freshwater/latin-america-water-funds.pdf.

5. "Africa's First Water Fund to Combat Rising Threats to Food Security, Water and Energy Supplies," Nature Conservancy, March 20, 2015, https://www.nature .org/cs/groups/webcontent/@web/@lakesrivers/documents/document/prd _288377.pdf.

6. Fred Pearce, "Poachers Turned Gamekeepers," CGIAR, *Thrive* (blog), July 8, 2015, https://wle.cgiar.org/thrive/2015/02/24/poachers-turned-gamekeepers.

40
Water Ethics

1. World Health Organization, "2.1 Billion People Lack Safe Drinking Water at Home," July 12, 2017, http://www.who.int/mediacentre/news/releases/2017/water -sanitation-hygiene/en.

2. Rural Water Supply Network, *Myths of the Rural Water Supply Sector*, RWSN Perspectives no. 4, 2010, available via Confama, https://www.comfama.com/contenidos /servicios/Gerenciasocial/html/Cursos/Johns-Hopkins/lecturas/Nagpal_Rural _Water_Supply_Sector.pdf.

3. Jamie Skinner, "Where Every Drop Counts: Tackling Rural Africa's Water Crisis," IIED briefing, 2009, http://pubs.iied.org/17055IIED.

4. Fred Pearce, "Doubts Raised over UN Drinking Water Claim," *New Scientist*, March 8, 2012, https://www.newscientist.com/article/dn21550.

5. Mark Sobsey et al., "Assessing the Microbial Quality of Improved Drinking Water Sources: Results from the Dominican Republic," *American Journal of Tropical Medicine and Hygiene* 90 (2014), https://doi.org/10.4269/ajtmh.13-0380.

6. Ameer Shahid et al., "Why 'Improved' Water Sources Are Not Always Safe," *WHO Bulletin*, 2014, http://www.who.int/bulletin/volumes/92/4/13-119594.pdf.

7. IWMI-Tata Water Policy Program, "Innovations in Groundwater Recharge," *Water Policy Briefing* 1 (2002), http://www.iwmi.cgiar.org/Publications/Water_Policy _Briefs/PDF/wpb01.pdf; Robert Ambroggi, "Underground Reservoirs to Control the Water Cycle," *Scientific American*, May 1977, https://www.scientificamerican .com/article/underground-reservoirs-to-control-t.

8. Natalie Greene, "The First Successful Case of the Rights of Nature Implementation in Ecuador," Global Alliance for the Rights of Nature, 2011, http://therightsof nature.org/first-ron-case-ecuador.

9. "India Court Gives Sacred Ganges and Yamuna Rivers Human Status," *BBC News*, March 21, 2017, http://www.bbc.co.uk/news/world-asia-india-39336284.

10. Kate Wheeling, "Rivers Win in the Fight for Non-Human Rights," *Pacific Standard*, March 30, 2017, https://psmag.com/rivers-win-in-the-fight-for-non-human -rights-bcd5e2141e08.

11. "Whanganui River in Pipiriki Expected to Reach over 13 Metres," Maori Television, April 5, 2017, https://www.maoritelevision.com/news/regional/whanganui -river-pipiriki-expected-reach-over-13-metres.

INDEX

Note: Page references in *italics* refer to illustrative matter.